The Ultimate Palm Robot

About the Authors

Kevin Mukhar is a programmer and writer. He has been programming as a hobbyist since the early days of the Commodore 64. He learned COBOL in high school and FORTRAN in college and has dabbled in everything from programming HP calculators to programming Java on the Palm OS. He is currently a software developer for Lockheed Martin in Denver.

Kevin has written numerous articles and is the co-author of several books, including *Beginning Java Databases* and *Beginning J2EE 1.4* (both published by Wrox Press). He served as Technical Editor for Dave Johnson's recent book *Robot Invasion: 7 Cool and Easy Robot Projects*. When not programming or writing, Kevin is working on a Masters degree in Computer Science, playing guitar, and trying to learn how to play saxophone.

Dave Johnson has a degree in Mechanical and Aerospace Engineering from Rutgers University, and he's been writing about technology for the past 15 years. He's the editor of *Mobility Magazine* and contributes to a number of magazines, including *PC World*, *Handheld Computing*, *Wired*, and *Tech Edge*. He's the author of more than two dozen books, including *How to Do Everything with Your Digital Camera*, and he recently wrote *Robot Invasion: 7 Cool and Easy Robot Projects*. Mostly, though, he's waiting for a call from Roger Waters to join him in a new rock band.

The Ultimate Palm Robot

Kevin Mukhar
Dave Johnson

McGraw-Hill/Osborne

New York Chicago San Francisco
Lisbon London Madrid Mexico City Milan
New Delhi San Juan Seoul Singapore Sydney Toronto

McGraw-Hill/Osborne
2100 Powell Street, 10th Floor
Emeryville, California 94608
U.S.A.

To arrange bulk purchase discounts for sales promotions, premiums, or fund-raisers, please contact **McGraw-Hill**/Osborne at the above address. For information on translations or book distributors outside the U.S.A., please see the International Contact Information page immediately following the index of this book.

The Ultimate Palm Robot

1234567890 CUS CUS 019876543

Book p/n 0-07-222882-2 and CD p/n 0-07-222881-4
parts of
ISBN 0-07-222880-6

Publisher
Brandon A. Nordin

Vice President & Associate Publisher
Scott Rogers

Acquisitions Editor
Margie McAneny

Project Editor
Monika Faltiss

Acquisitions Coordinator
Tana Allen

Technical Editor
Steve Richards

Copy Editors
Judith Brown, Lisa Theobald, Lunaea Weatherstone

Proofreader
Claire Splan

Indexer
Irv Hershman

Composition
Tabitha M. Cagan, Lucie Ericksen

Illustrators
Lyssa Wald, Kathleen Fay Edwards, Melinda Moore Lytle, Michael Mueller

Cover Design
Pattie Lee

Cover Photograph
Dave Bishop

This book was composed with Corel VENTURA™ Publisher.

For my wife, Anne.
—Kevin

For the two greatest kids on the planet, Evan and Marin.
—Dave

Contents

Acknowledgments

Any time you write a book that is equal parts research, construction, writing, and tech editing, you know you have a long and arduous road ahead. Trust us: there's a special place in hell for anyone who says, "Turn a Palm into a robot? Then write a book about it? How hard could that be?" That said, the stellar folks at McGraw-Hill/Osborne made it as painless as possible, and we thank them heartily for that. We'd like to thank Margie, Monika, and Tana from the bottom of our hearts for keeping this book on track and making it look as good as it does.

We also want to thank the Manipulation Lab, of the Robotics Institute at Carnegie Mellon University, for the original vision, design, and construction of the Palm Pilot Robot Kit. Thanks to the folks at Acroname who took the next step in creating a Palm Pilot Robot suitable for hobbyists. Special thanks to Steve at Acroname for the help he provided in understanding some of the idiosyncrasies of the robot. Finally, thanks to all the hobbyists and programmers who let us use their software in this book.

Kevin would like to thank Dave, who has been more than a just co-author. Thanks, Dave, for asking me to write this book with you, and thanks for being there when I needed your help. Thanks also to my family for letting me disappear from their lives for the last four months.

Dave would like to add that Kevin is a great friend and co-author. Thanks, Kevin, for teaming up with me and losing four months of your life to this project. I know you shouldered the majority of this book, and it's something to be proud of. Also, thanks to my wonderful family and, as always, my cat Hobbes, without whom I would have had significantly greater mobility throughout the writing process.

Introduction

As kids who grew up in the '60s and '70s, we were excited by the prospect of robotics. Shows like *The Jetsons* and movies like *Forbidden Planet* were hints of the fantastic possibilities of a world filled with robots.

Unfortunately, the 21st century has arrived and robots are still, in many ways, distant dreams. Upon hearing that the army was using robots in Afghanistan last year to root out mines, we were a bit disappointed to see that they were little more than Battle Bot–like remote controlled cars. Sheesh.

On the bright side, consumer technology is converging with bleeding edge technology in ways unimaginable a decade ago. If real robots look like soup cans with wheels and antennas, then it stands to reason that ordinary folk can make their own robots that are almost as sophisticated, right?

Right! That's the motivation for this book, which we hope will excite and inspire you to launch your own career in robotics, even if it's only in your basement. Equipped with a Palm OS PDA and some inexpensive parts, almost anyone can create a fully functional robot. Heck, we did it. And so can you.

While we intend this book to be a resource for beginners as well as experienced hobbyists, it is geared more toward the beginner. For that reason, we begin in Chapter 1, *Meeting the Palm Robot*, with a gentle introduction to the world of robots. We'll look at what makes a robot, some of the various types of robots, and why the Palm Pilot Robot is ideal for beginners.

Using a kit to build something is always easier than doing it yourself. In Chapter 2, *Getting the Parts*, we'll show you where to get a Palm Pilot Robot Kit, or, if you want the challenge, which parts you need to get to do it yourself. In Chapter 3, *Building the Robot*, we'll show you how to put it all together. There are two versions of the robot; we'll show you how to build the original PPRK using the Pontech SV203 controller, and how to build the newer model that uses the Acroname BrainStem controller. For a few Palm OS devices, you'll need to do some additional work on your robot for it to function with the Palm OS device. Chapter 4, *Using Palm VIIs and Handspring Visors*, shows you what you need to do to get the robot to work with Palm VIIs and Handspring Visors.

After you've built the robot, you need to get software to make it do something, such as chase objects, follow patterns, or conquer the world. In Chapter 5, *Checking Out the Robot*, we'll show you where to get software developed by other hobbyists, and how to use that software to test the operation of your robot. The real fun starts when you can develop your own software and watch your robot do what you want it to do. In Chapter 6, *The Palm Robot Programmer (PRP)*, we introduce an application that we developed specifically for this book. Using the Palm Robot Programmer, you click icons and use those icons to create a program for your robot. When the program is complete, the Palm Robot Programmer will convert the program into computer source code for you. Chapter 6 will also provide a brief introduction to how to use the code with your robot. While the Palm Robot Programmer is meant for those new to programming, experienced programmers will want to check it out as well. The Palm Robot Programmer is designed so that experienced programmers can extend it for new programming languages.

Experienced programmers will also want to check out Chapter 7, *Essential Robot Programming Strategies*, and Chapter 8, *Taking Control of the BrainStem Robot*. In Chapter 7 we show you how to write your own programs for the Pontech SV203 controller. We look at programming in BASIC and C, and write a simple program in both languages that will show you the basics of how to control the robot. In Chapter 8, we look at how to write programs for the other robot controller, the BrainStem. In this chapter, we'll look at writing programs in a language named TEA and in Java.

In Chapter 9, *Sensors and Enhancements*, we turn back to the hardware side with a look at other sensors and devices that you can attach to your robot. We'll show you how to connect them and how to use them.

As you probably know, you can do a million and one things with your Palm OS PDA. In Chapter 10, *Having Fun with Your Robot*, we turn away from the Palm Pilot Robot and look at a few other robot-related activities that you can do with your Palm OS PDA.

Finally, several of the chapters in this book use software tools to write programs for the robot. We've gathered all the software that we used or developed in the book, including the Palm Robot Programmer, and put it all on a CD-ROM.

Chapter 1

Meeting the Palm Robot

There's nothing quite like robotics. Sure, there are all sorts of cool, fun, and exciting technologies out there, but only robots can so thoroughly ignite the imagination. Maybe it's because robots mimic life. Making a robot is like breathing life into a collection of metal parts and circuit boards. Perhaps we desperately want mechanized slaves to do our bidding, like Robbie from the movie *Forbidden Planet*. Maybe robots have simply been associated with "the future" since the 19th century, and they represent our possible utopian world.

Whatever the reason, people have been tinkering with home-grown robots for as long as scientists and engineers have been making them in the laboratory. In this book, we'll help you turn a common PDA—your old Palm OS handheld—into a simple robot. Ready? Before we begin, let's take a quick look at the world of robotics.

What Is a Robot, Exactly?

With so many sorts of robots—in movies, working with emergency crews, in our living rooms, and on drawing boards—you might begin to wonder what, exactly, the definition of a robot is. After all, it seems that radically different devices all go by the same name. Do robots have to be autonomous? Can you build a radio-controlled device and call it a robot? Where do you draw the line?

For an answer, let's turn to the dictionary. The American Heritage Dictionary defines a robot this way:

> *An externally manlike mechanical device capable of performing human tasks or behaving in a human manner.*

Okay, that'll work, but it's hardly a complete definition. If we go strictly by this definition, then in order to be a robot, a device would essentially have to look or act like a human. We've all seen robots that look like cars, planes, bugs, trash compactors, and dogs. In the 1970s, the Robot Institute of America established a far more complete definition of a robot:

> *A robot is a reprogrammable, multifunction manipulator designed to move material, parts, tools, or specialized devices through various programmed motions for the performance of a variety of tasks.*

That definition works a lot better, since it covers almost all of the robots you commonly (and don't commonly) see or hear about.

Asimov's Three Laws of Robotics

Most sci-fi fans are familiar with Isaac Asimov's three laws of robotics. Designed as a simple narrative device to help tell the story of a robot on trial for killing a human, the three laws have become an essential part of the science fiction landscape, co-opted by dozens of writers, television shows, and movies.

The laws made their first appearance in Asimov's classic 1942 story *Runaround*, and were later incorporated into a series of robot-centric stories and novels. Here are the three laws as Asimov first wrote them:

1. A robot may not injure a human being, or, through inaction, allow a human being to come to harm.
2. A robot must obey orders given it by human beings, except where such orders would conflict with the First Law.
3. A robot must protect its own existence as long as such protection does not conflict with the First or Second Law.

The Essence of a Robot

Even though robots vary dramatically in appearance, you can generally characterize them as being made of four key components:

- **Base** The base is the body of the robot. In our Palm Robot, the base is a simple metal chassis on which the wheels, motors, and Palm PDA itself are mounted. Not all robot bodies need to support motion, but ours will.
- **Processor** Our processor will be a Palm OS PDA. Most Palm-powered PDAs use a Motorola DragonBall processor, so that's the heart of our bot.
- **Actuators** Actuators are the muscles that make your robot interact with its environment. We'll use a trio of motors to power the wheels that make the Palm Robot roll.
- **Sensors** Sensors are the eyes, ears, and nose of the robot. Not all robots have sensors; if a robot is sufficiently simple, it doesn't need to measure its environment in order to carry out

its programming. A simple moving robot might sense edges, colors, shapes, or paths in order to stay on the right track, though, and some robots can use temperature, light, motion, or odor sensors in order to carry out their job. We'll talk in Chapter 8 about possible sensor enhancements you can add to your Palm Robot.

The Wide, Wide Worlds of Robots

For the last 20 years or so, researchers and hobbyists have been developing robots and pushing forward the boundaries of this fledgling science. At the university level, small-scale research projects abound. Here we'll tell you about some of the most interesting robots.

BEAM Robotics

BEAM stands for biology, electronics, aesthetics, and mechanics. The behavior of BEAM robots is driven by their electronic circuitry, with no programming required. They're intended to be autonomous; that is, once activated, they should be able to operate for relatively long periods of time without outside intervention. For that reason, they are usually solar-powered devices. Some of the earliest work in BEAM robotics was done at Massachusetts Institute of Technology. You can read more about these robots at www.ai.mit.edu/people/brooks/projects.shtml.

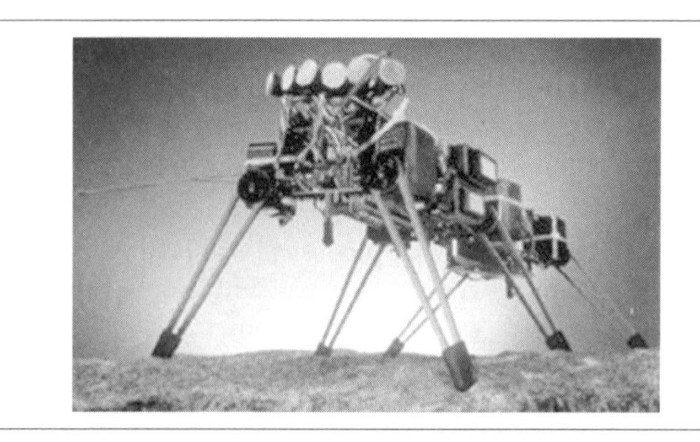

Stiquito

The Stiquito is a great example of simplified robotics. The Stiquito is not programmed. Like BEAM bots, its behavior is driven by the electronic circuitry that the robot is constructed from. (The Stiquito is not a BEAM robot because it is battery powered, and thus not able to act autonomously for long.) With the Stiquito, the electronic circuit passes current through the legs of the robot; the wire legs are made of a special material called Nitinol, also known commonly as a "muscle wire." Nitinol contracts when current heats the wire, and expands when cooled, thus simulating the way a muscle works in a living creature. Stiquito was originally developed at Indiana University (www.cs.indiana.edu/robotics/stiquito.html), and you can buy inexpensive Stiquito kits to make your own.

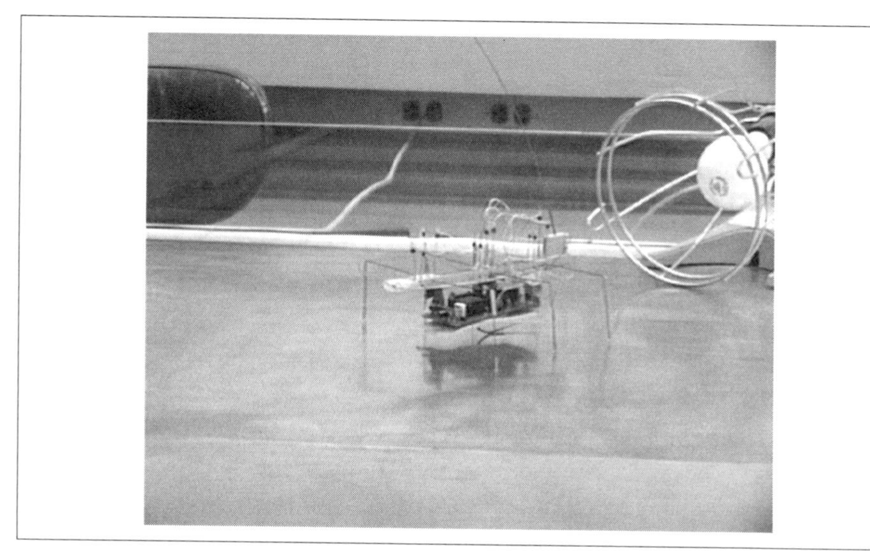

Project Timber

The majority of robots rely on programming, not prewired circuits, to drive behavior. Project Timber is a research program of the Computer Science and Engineering department at the Oregon Health and Science University. The robot in this project is known as Timbot, which is short for Timber Robot. The brain of the robot is a Pentium III CPU that is mounted in a radio-controlled monster truck model. The Timbot is an example of a robot that puts all the intelligence of the robot into a computer that controls the robot's behavior. Unfortunately, for hobbyists like us, the programming aspect is rather formidable.

Not only does it need to be programmed, its designers developed their own unique programming language for the robot. Read more at www.cse.ogi.edu/PacSoft/projects/Timber.

The Palm Pilot Robot Kit

The Palm Pilot Robot Kit (PPRK) was developed at Carnegie Mellon University (www-2.cs.cmu.edu/~pprk). One of the design goals was to create a robot that anyone could build. The body of the robot is made of easily obtained materials, and construction requires just tape, glue, and a little soldering. For its brain, the original PPRK used a first-generation Palm Pilot. Why? PDAs like the Palm are small, have a very long battery life, use a graphical display, and are relatively easy to program. This makes the PPRK ideal for hobbyists who want to experiment with more complex robots. The PPRK design relies on three wheels rather than four. Because of the unique design of the wheels, the robot can rotate in place and can move in a straight line.

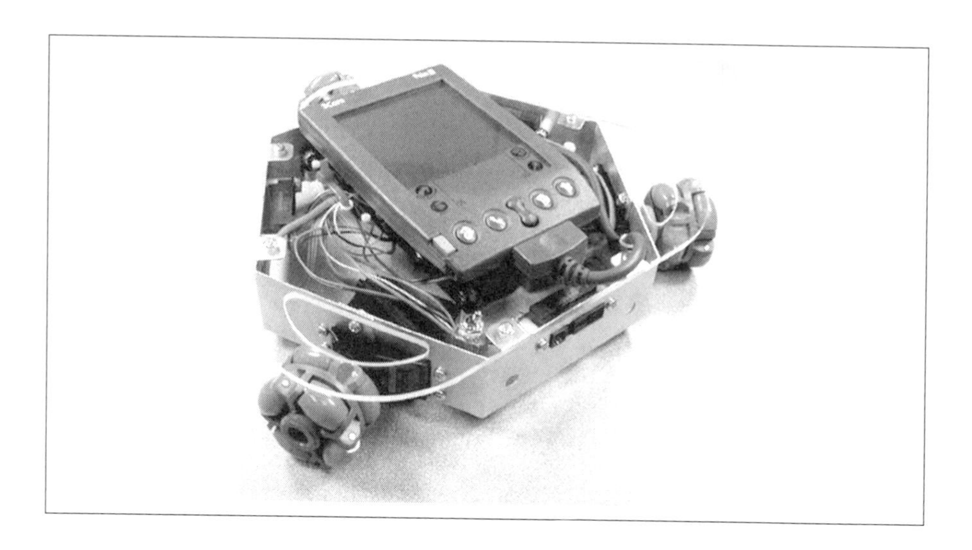

The PPRK

So, we're guessing that unless you bought this book by mistake, you are reading it to build your own Palm Pilot Robot Kit. The PPRK is easy to complete, but the online instructions leave a lot to be desired—thus, this books fills in the gaps.

Like the *Blair Witch Project*, *Memento*, and Pauly Shore's acting career, PPRK was something of an unexpected success. It proved so successful, in fact, that Carnegie Mellon licensed the design to a company called Acroname, Inc., for marketing purposes. The end result? If you're a casual hobbyist, you can get all the parts you need to build a Palm OS robot from one location. No need to scour the bins of your local electronics store looking for the right servo, or the correct wheels. Acroname has even developed a model that does not require a Palm OS PDA to be attached to the robot. Of course, you can still go bin scouring if you want to; most of the parts for the PPRK can be easily obtained from hardware and electronics stores.

While the kit to re-create complicated robots may be easily accessible to the casual hobbyist, the programming probably is not. For example, with the

Timbot that we mentioned earlier in the chapter, you need to learn a new programming language to be able to use the robot. If you're not a programmer, you need to learn how to program as well. We're willing to bet that you are more interested in building a robot that does interesting things, rather than learning how to program.

But doesn't the Palm need to be programmed?

Well, yes. But if you're the hobbyist who wants to spend time building and playing with the robot rather than learning how to program, this book will come in handy. We'll show you where to go on the web to get programs you can use with your robot. We'll also provide you with a simple application that that you can use to create your own programs; it does most of the programming work for you.

And if you want to do your own programming, you'll get the chance to sling code to your heart's content. Programming for the Palm uses existing languages like C or BASIC that many programmers already know. So if you already know how to program in C, C++, BASIC, or Java, this book will get you started.

Before we get to the robot itself, let's take a quick look at the Palm PDA.

The Rise (but Not Fall) of the Palm Empire

The Palm was not the first PDA to hit the streets, but it certainly was the first successful one. One of the first PDAs was the Apple Newton. Dave, gadget freak that he is, owned a Newton. Okay, he owned several of them. But we digress.

The Newton was an extremely capable device, but it had a few fatal flaws, the most important of which was its size. Unlike today's PDAs, you couldn't fit a Newton into your pocket. Well, not unless you had really big pockets. And it was relatively heavy, so if you did put it into your pants pocket, your pants would sag halfway down to your knees. And thus the Newton led to that great fashion statement for young men of the nineties: droopy oversized pants with big pockets.

The world of PDAs changed drastically in 1994, when Jeff Hawkins, a developer of software for the Newton and other fledgling handheld devices, founded a company called Palm Computing and decided that he could make a better PDA. Hawkins envisioned a pocket-size computer that would organize calendars and contacts, and perhaps let travelers retrieve email while on the road.

Hawkins knew he'd have a tough time selling his device, so he decided to convince himself before trying to convince investors. His device would be roughly the size of a deck of cards—much smaller and lighter than the Newton—and would therefore fit in a shirt pocket. But would it be practical even at that size? Would it be comfortable to carry around? Hawkins decided to find out. Before a single piece of plastic was molded, before a single circuit board was designed, the Palm Computing Pilot existed solely as a block of wood.

Hawkins cut a piece of balsa wood to the size he'd envisioned for his handheld device, put it in his shirt pocket, and left it there—for several months. He even took it out from time to time and pretended to take notes, just to see if the size and shape felt right. Though he quickly came to realize that such a form factor made perfect sense, doors slammed whenever he showed the "product" to potential investors. "The handheld market is dead," was the mantra at the time.

Fortunately, modem-maker U.S. Robotics liked the idea of the Pilot so much that it bought Palm Computing outright. In March 1996, the company unveiled the Pilot 1000, and the rest is history.

Flash forward seven years. The Pilot—which would go through a number of name changes, first to PalmPilot and then just Palm—had become the fastest-growing computer platform in history, reaching the million-sold mark faster than the IBM PC or Apple Macintosh. In the interim, U.S. Robotics had been assimilated into networking giant 3Com, and Palm along with it. The Palm line had grown to include a variety of models, and companies like Handspring, IBM, and Sony had adopted the Palm operating system for their own handheld devices. Recently, Palm announced that it had sold its 20 millionth PDA—not bad for a company that started out based on a little block of wood.

And thus, the first Palm devices, the Pilot 1000 and Pilot 5000, were born. These were followed by the PalmPilot Personal and Professional. Due to a dispute with the Pilot company, models that followed the Professional were just called Palms.

The Palm: A Good Robot Brain

At the end of *The Wizard of Oz*, the wizard gives the Scarecrow a degree, and suddenly the Scarecrow begins thinking great thoughts. It's a nice idea, but of course that won't work for our robot. No matter how many degrees we buy from those late-night TV infomercials, our robot still needs a brain. (But wait folks, there's more. If you buy your degree today, we'll throw in, absolutely free, a fabulous set of kitchen knives!)

So, since the degree idea won't work, we need to find a good brain for our Palm Robot. There are lots of digital devices to choose from: personal organizers, mobile phones, cameras, digital toothbrushes, and the list goes on. Okay, digital toothbrushes don't exist, but if they did, Dave would have one, but we still wouldn't be able to use it as the brains for a robot.

Our robot brain must meet certain requirements.

- ❑ It must be small enough to fit into the PPRK.
- ❑ It must be able to store and execute a program that controls the robot.
- ❑ It must be able to communicate with the robot.
- ❑ It must be easy to program.

Does the Palm fit that criteria? Absolutely. Let's break it down, item by item.

Size

PDAs are the perfect size for our robot. There are smaller robots, where the size of even a PDA would be too large, and there are bigger robots in which size wouldn't matter (although it is still easier to mount a PDA than a PC on a robot). But the electronics used by the PPRK, and the size of the servos and wheels, make the PPRK just a little larger than the Palm itself. Unlike aliens in science fiction whose brains are larger than their bodies, most living things on earth have bodies that are larger than their brains. The same is the case for our robot. Trying to put a PC on the chassis of the PPRK would be too difficult, but the Palm fits just right.

CPU

We keep saying that the Palm is the brain of the robot, but really, the brain is the CPU that powers the Palm. As we mentioned earlier, the CPU inside the vast majority of Palm OS devices is Motorola's DragonBall processor. It's not particularly fast or otherwise advanced, but the DragonBall is a CPU very much like the one that powers your desktop computer. It does all the same functions, input, processing, and output, just like a desktop computer.

This is important for us because the computer programs that drive the robot will need to be processed by this CPU. Dave's digital toothbrush wouldn't work well because it doesn't need a full-featured CPU. The Palm, on the other hand, does. While embedded processors inside most digital devices tend to have very little functionality (to save cost), or are quite specialized to cater to the limited needs of the device, the Palm's DragonBall is a general-purpose processor, thus more suited for use in a robot that needs to be programmed for many different tasks. As an added bonus, most Palms come with plenty of memory for the programs we will put onto the PDA.

Communication

The brain of our robot needs some way to communicate with the electronics and motors that power the robot. PDAs are perfect for this. Although PDAs can function without a PC (and certainly, some people own PDAs without owning a PC), they are designed to be able to connect to a PC. Why? Most PDAs are considered to be companion products to the PC, sharing and synchronizing data with the programs you already have on the desktop. Plus, the PC is a convenient conduit for installing new programs on your PDA. This communication for the Palm originally occurred with a standard serial cable (although it had a special connection on the Palm end). The communication protocol was well documented so that developers could write their own programs that transferred data to and from the PC.

It is this communication channel that makes the Palm perfect for controlling our robot. As we will see in later chapters, the controller for the robot communicates with the PDA through a serial cable, which attaches to the PDA's standard synchronization port (see Figure 1-1). Readings from the sensors attached to the robot are sent to the controller, which passes them over the serial cable to the PDA; likewise, the PDA can determine which servos to actuate and send the command to the controller, which routes the command to the correct servo.

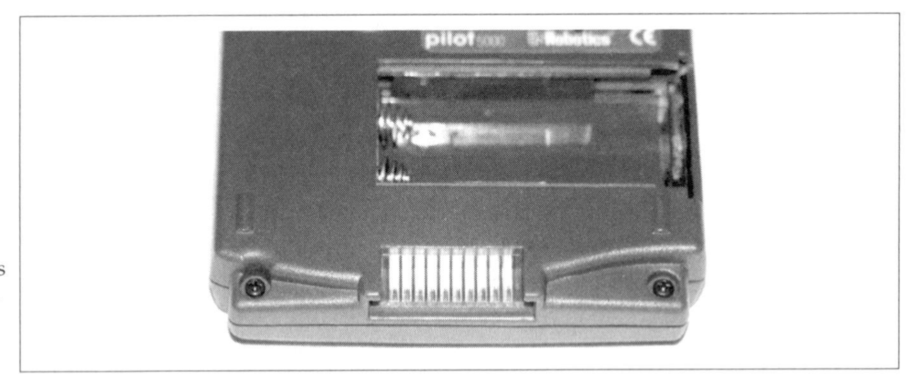

Figure 1-1 The Palm's serial connector is a logical way for the CPU to interact with the sensors and actuators on our robot.

Easy to Program

The programming for the Palm was originally done in the C language, though there are now several less complex alternatives available, including a BASIC interpreter and a Java Virtual Machine. Thus, you don't need to learn a special language to program the Palm. If you know C or C++, BASIC, or Java, you can program the Palm PDA.

And, as we mentioned earlier, if you don't know any of those languages, this book will give you a tool that you can use to create programs without needing to know how to program. The application is included on the CD for this book (along with other programming tools if you want to program the Palm directly). When you're ready, be sure to flip over to Chapter 6 for all the details on this.

Palm OS Devices from A to Z (or 1000 to VIIx)

The PPRK from Acroname can be run with almost any Palm OS PDA ever created. It's possible that you may even have one of these devices hanging around. The list of Palm PDAs that follows is by no means complete; primarily we'll look at the Palm OS PDAs that are compatible with the PPRK. So join us as we take a trip down memory lane, looking at some Palm OS PDAs past and present.

The Pilot 1000 and Pilot 5000

The very first Palm OS devices, released in 1996, were the Pilot 1000 and Pilot 5000 (see Figure 1-2). If you had one of these devices, you had either 128KB (the Pilot 1000) or 512KB (the Pilot 5000) of memory. These devices included early versions of the Palm Address Book and Date Book, a To Do list, and a Memo Pad. You might be able to find one of these around, and you may even be able to use it with the PPRK, but since the first PPRK used a later Palm model, we won't talk much more about these.

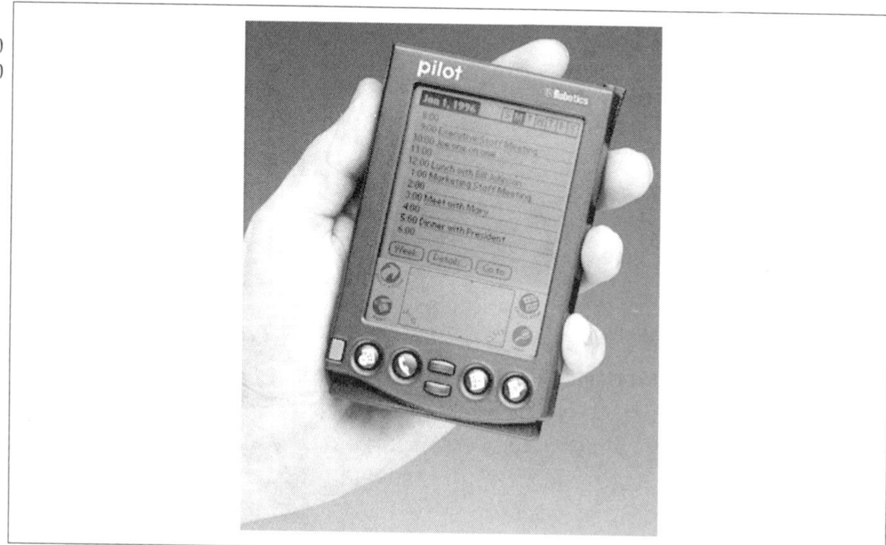

Figure 1-2 The Pilot 1000 and Pilot 5000 are a piece of PDA history; they were the first models sold.

PalmPilot Personal/Professional

The PalmPilot Personal and the PalmPilot Professional, shown in Figure 1-3, were released in 1997. The Personal featured 512KB of memory, and the Professional had a whopping 1MB of memory. Even more useful, though, is that both introduced backlighting, which made them usable in low light situations. The other important feature introduced by the Professional was a mail application. Initially, you had to download mail to your PC first, and then transfer it to your PDA during a HotSync. Later, Palm offered a modem that let you connect to the email server and download directly to the PDA.

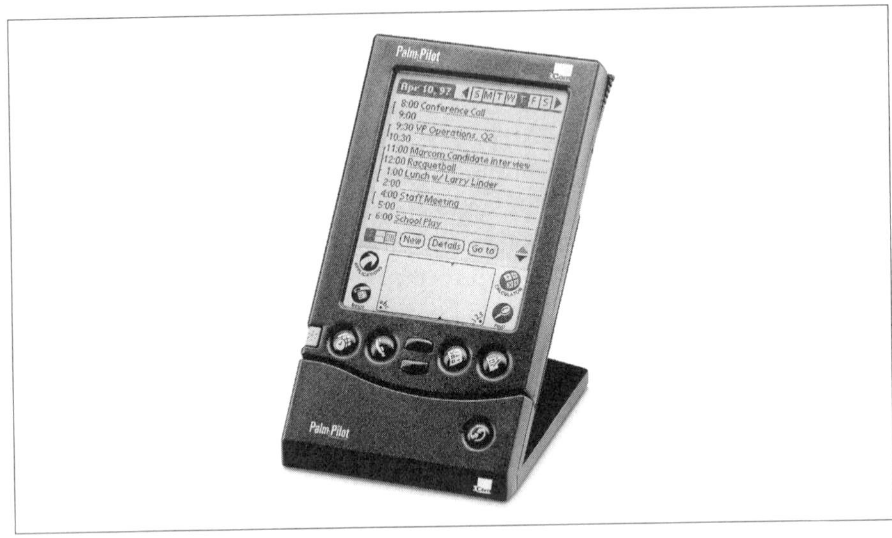

Figure 1-3 The PalmPilot Personal and PalmPilot Professional were even more popular than their predecessors.

Palm III

In 1998, the Palm III was released. This model became one of the most popular and longest-lived PDAs of all time. One reason for its popularity is that Palm released four variations on the same basic platform—including the first Palm with a color screen. (See Figure 1-4 for a look at the color Palm.) The Palm's infrared port also made exchanging data between devices very easy. While no longer sold, there are millions of Palm III models still in routine use. That makes this model one of the easiest to find on the used market.

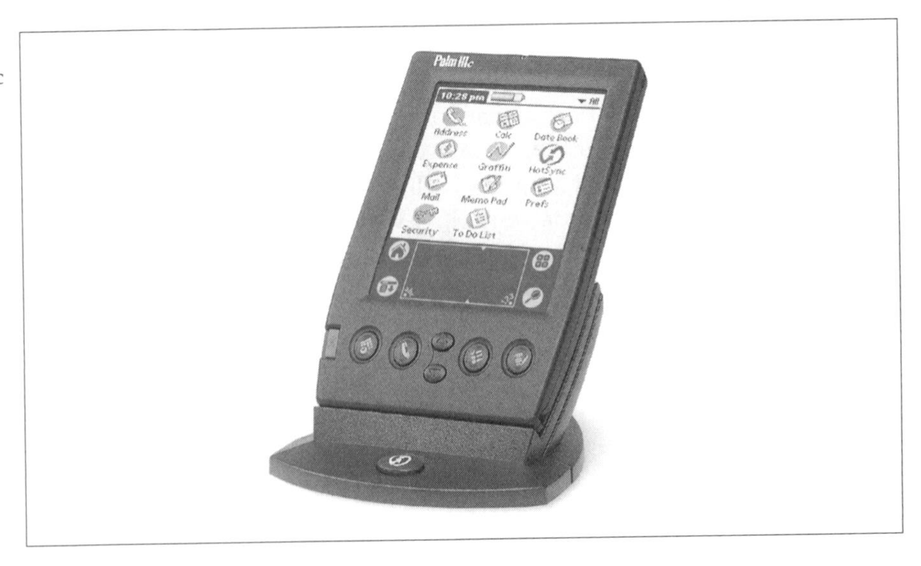

Figure 1-4 The Palm IIIc was the first color device offered by Palm.

Palm V

How do you follow up such a runaway hit as the Palm III family? By adding an integrated, rechargeable battery that tops off whenever the PDA is sitting in the cradle. The Palm V also sported a sexy, brushed aluminum case design (check out Figure 1-5) that made it incompatible with earlier docking cradles, but made it a hit with executives around the world. The Palm V is also quite easy to find on the used market thanks to its immense popularity.

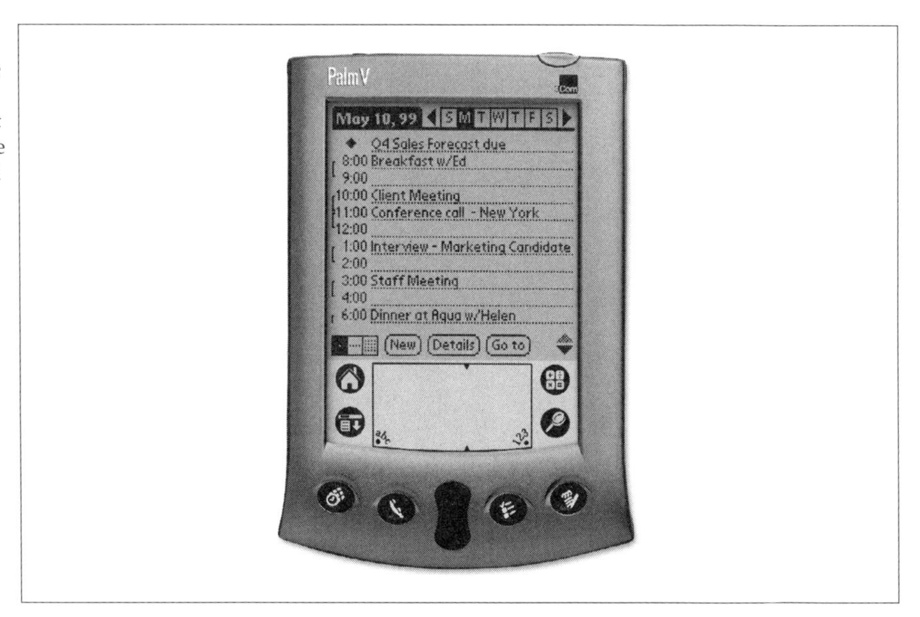

Figure 1-5 Some believe the Palm V was the most stylish device ever released by Palm.

Palm VII

The Palm VII came to the market in 1999. The end result of a long development program to deliver wireless email and limited web surfing in a handheld device, the Palm VII never proved to be very popular. It was eventually replaced by the i705 in 2002. In its heyday, though, it was essentially a Palm III with an integrated wireless modem. See Figure 1-6 for a look at the Palm VII with its antenna extended.

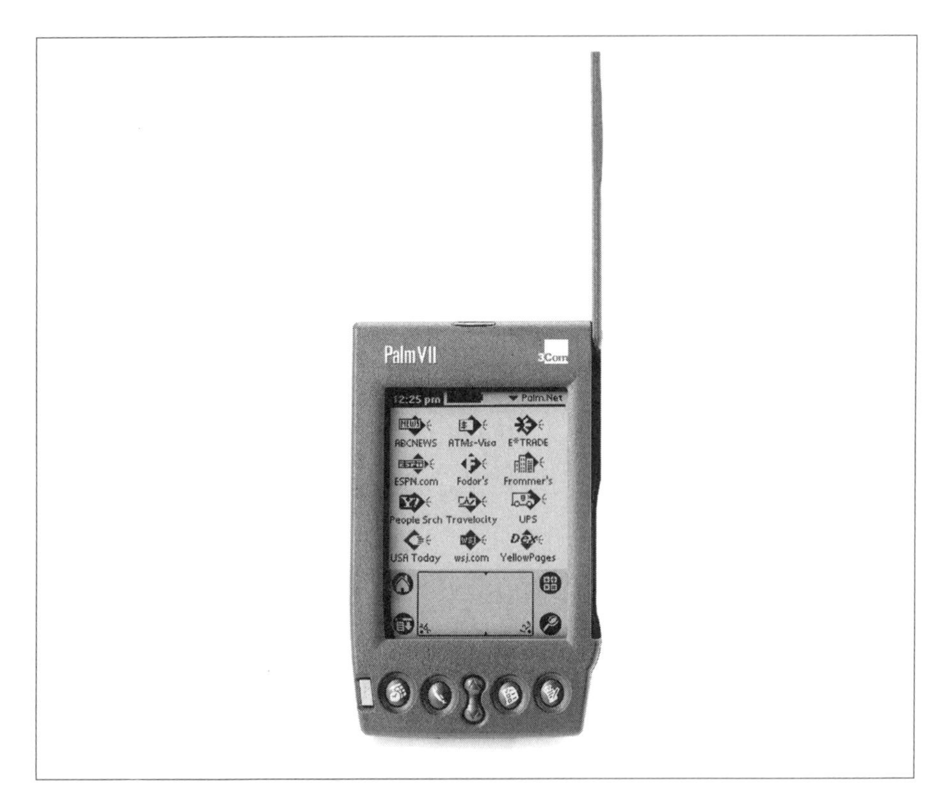

Figure 1-6 Here you can see the Palm VII with its antenna extended.

Other Palm Models

Since the Palm VII, Palm has continued to innovate and introduce a number of new models. These days, store shelves are only stocked with these newest models. These include the many models in the *m* series—from the m105 to the m515—along with the Zire, the i705, and the Tungsten T and Tungsten W. One of the most affordable Palm devices to date is the m105, seen in Figure 1-7.

At the time this book was written, the PPRK did not support any of these latest models in the Palm family, though that may change by the time you read this. Acroname has assured us that it is working on a cable that is compatible with the many devices in the *m* series. The main problem for the m105 and other *m* series models is that the current connector cable for the PPRK uses a Palm III–style connection; this connector can connect to the m105, but does not stay firmly attached. As a result, you need to glue or otherwise attach it to the m105, which is problematic if you want to use your m105 for things other than robotics.

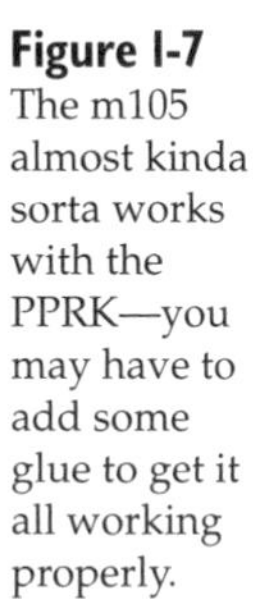

Figure 1-7 The m105 almost kinda sorta works with the PPRK—you may have to add some glue to get it all working properly.

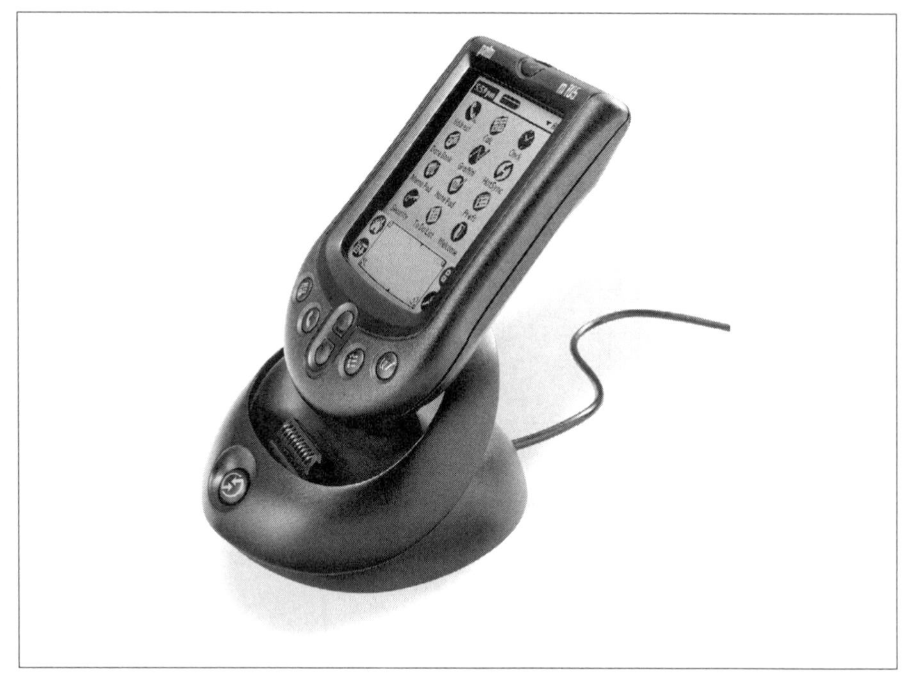

Handspring Visor

A few pages back, we told you the once-upon-a-time story of Jeff Hawkins and his bid to make the perfect PDA. A few years after Palm launched its first few devices, Hawkins left Palm and started a new company called Handspring. Hawkins' goal in creating Handspring was to create *smartphones*—PDAs that doubled as mobile phones. It took a few years for the first Handspring smartphone to materialize—you may know it now as the Treo—but the company's first few devices were Palm OS devices that went by the name Visor (see Figure 1-8).

The Visor was similar to the Palm III, but it featured a key innovation: an expansion slot in back called Springboard. Although not important for the PPRK, Springboard proved to be important to Handspring for several years, since it allowed dozens of independent developers to create add-on modules for the Visor. These gadgets turned the simple Visor into GPS receivers, FM radios, cell phones, digital cameras, and more.

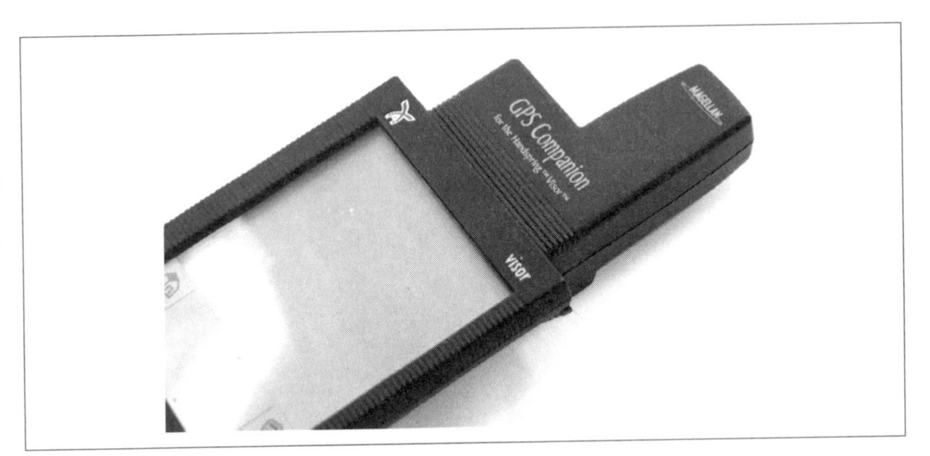

Figure 1-8 The Handspring Visor looks much like a Palm III, but features the Springboard expansion slot in back.

The Robot Geek Says

The Visor and the PPRK

- If you are planning to use a Visor with the PPRK, we strongly recommend you get the BrainStem model from Acroname (www.acroname.com).
- You can use the Visor with the original PPRK, but it involves a lot more effort and requires you to modify the controller circuit board.
- Using a Visor with the original PPRK is definitely not a project for the casual hobbyist.

The Other Guys: Pocket PC

Those of you readers who are Palm fanatics should probably move on to the next section right now. That's right… move along… nothing to see here.

Okay, for those of you who are left, we would be remiss if we didn't mention the other major class of handhelds—Pocket PCs. Let's be clear: the PPRK kits and this book only cover making a robot with a Palm OS PDA, or running stand-alone without any PDA. However, there's no technical reason you couldn't use a Pocket PC device—running Microsoft's Pocket PC operating system—instead. As we'll see later, Acroname, the company that sells the PPRK, provides lots of software and tools for using a Pocket PC with the PPRK.

We use the term "Pocket PC" for any device that runs the Windows Pocket PC operating system. Pocket PCs are sold by a number of companies, including Hewlett-Packard, Toshiba, Dell, ViewSonic, and others (see Figure 1-9).

Figure 1-9 Pocket PCs, regardless of how they're branded, can all run the same software and, in fact, all work more or less the same way.

Since Pocket PC devices run a version of Windows, you may be familiar with many of the programs for these devices. Many Pocket PC programs are smaller scale, "pocket" versions of the same programs that run on a full-sized PC running a Windows OS.

Where to Find New and Used PDAs

So, you've decided to build the Palm Robot. (If you're still reading this in the aisle of your local bookstore, now's the time to decide. Stop procrastinating, take this book to the checkout, and buy it.)

Obviously, you will need to get yourself the parts for the robot. We'll talk about robot parts in Chapters 2 and 3. But the other important ingredient is a Palm OS PDA. How can you find a cheap PDA that will be no great loss if it gets banged up a bit as part of a robot? Let us show you.

The most obvious place to get a PDA is to reach down to your belt and grab your Palm out of its carrying case. If you already own a PDA, and you don't mind devoting it to your robot—at least part of the time—then there's nothing else you need to do. Move on to the next chapter.

If, on the other hand, you don't own a PDA and your hand comes up empty when you reach down to your belt, you'll need to get one (a PDA that is, not a belt).

You can obviously run down to your local electronics store and buy a brand new PDA. That, however, may set you back hundreds of dollars for the very latest PDA gadget. If you want to take a more economical approach, there are various ways to get your hands on pre-owned PDAs. We present a few of them here.

Should You Buy Used?

Just how reliable are those used PDAs? Should you invest in a PDA that's been used and possibly abused for years before you got your own hands on it?

Consider this: any time you buy something secondhand, it has undergone a certain amount of wear and tear. That means any used PDA may have certain defects, such as buttons that don't respond as well, components that are closer to their end of life, and rechargeable batteries that have measurably less capacity than when the device was new. A particular concern with used PDAs is the state of the screen and Graffiti area. Since people interact directly with those parts of the PDA, they can get scratched and sometimes even damaged. A few scuffs is no big deal, but an abusive owner can really do some damage to a PDA touch screen.

One thing is for sure: with 10–15 million PDAs sold each year, there's a big secondhand market. If you can get a PDA inexpensively enough, a little wear is no big deal.

Your Friend the Gadget Freak

Do you have any friends who are gadget freaks? (Not sure? You can identify them by the way their clothes sag due to all the gadgets they carry.) These folks were first to get a cell phone, first to get the latest gaming console, and right now they're standing in line at the theater so they can be first to see the *Matrix* sequel. That's probably where Dave is right now.

If your friend bought a PalmPilot Professional when it first came out, upgraded to the Palm III, followed by the Palm V, and then the m515 as soon as they were available, this person might still have his older Palm devices lying around. Ask if you can borrow, or even have, one of the older PDAs for use with a cool robot project. Who knows, your friend may even want to help build the robot.

The Local Recycled Electronics Store

If you live in a fairly large city, chances are good that there's an electronics store that recycles computer parts. This store will have old computer cases, motherboards, hard drives, 5 ¼-inch floppy drives, 8-inch floppies (try telling a teenager today that there used to be 8-inch floppies; they won't believe you), cables, wires, resistors, diodes, and, if you're lucky, old PalmPilots. While you're at it, you may also find parts for your robot.

Pawnshops

Pawnshops do tend to have a less than stellar reputation, but they have their uses—Dave bought a guitar at one many years ago, though Kevin refuses to go near any such establishment. However, they are one place that tends to have used electronic items. If there's a pawnshop in your area, you may want to consider checking for any used PDAs.

Online Sellers

Various online retailers sell digital devices. You can, of course, find current devices at retail prices. There are also places online to find older, cheaper devices. Here are a few places where you can find PDAs online.

eBay and Yahoo Auctions

One insanely popular place you can look for used Palm OS devices is an online marketplace like eBay or Yahoo auctions. eBay is a treasure trove of used electronic gadgets. When we recently checked for PDAs, we performed a simple search for "Palm pilot" and got over 1000 hits. When searching for "handspring visor," over 1300 listings matched the search. Not all of these are the PDAs themselves; a lot of the listings were for accessories such as cables, keyboards, cases, and other gadgets. But a lot of those hits were PDAs, and that means you can get them at good prices.

Other Online Sources

There are many other online sources for PDAs. You can perform a search for PDA resellers, but a better option is to surf news and message boards dedicated to PDAs. Try these:

- www.pdabuzz.com
- www.brighthand.com
- www.palminfocenter.com
- www.cliesource.com
- www.the-gadgeteer.com
- www.bargainpda.com

The folks at any of those sites can probably suggest good places to get used PDAs.

Stuff to Come

By now you're probably ready to move on to the real stuff—like how to build your robot, how to program your robot, how to do cool stuff with your robot, and how to conquer the universe with your robot.

We can't help with that last one, and frankly we don't think you'll get very far. People who try to conquer the universe with robots are usually stopped by Dr. Who, occasionally by Jim Kirk, and sometimes by the Interstellar Alliance created by the former commander of Babylon 5.

We can help with the first three.

In Chapters 2 through 4, we'll look at all the hardware you need to build the Palm Robot. As we've mentioned, Acroname sells some kits that have everything you need to build the robot. If you're a hard-core hobbyist, you may want to do it more from scratch. We'll show you the parts you need and tell you some places where you can get the parts.

In Chapters 5 through 8, we'll cover the programming aspects of the Palm Robot. In Chapter 6 we show you the Palm Robot Development Environment. This is the software we developed that allows you to program the Palm Robot without knowing any programming. If you want to sling code, we show you how to do that in Chapter 7.

In the final chapter, Chapter 9, we'll look at some other cool robotic applications for Palm PDAs.

So grab your soldering irons and screwdrivers, it's time to start building your Palm Robot.

Chapter 2

Getting the Parts

There are two approaches to building the robot—the right way and the wrong way.

Just kidding: you can build your Palm Robot from a kit or build it from scratch, and either approach is fine. Consider this: if you are more interested in playing with your robot than in dealing with all the nuts and bolts, or if you are a beginner when it comes to electronics, then you will probably want to build the robot from a kit.

On the other hand, if you're an adventurous beginner or an experienced hobbyist, then you may want to do more of the work yourself. In that case, you will want to get the parts yourself rather than relying on a kit.

In this chapter, we'll talk about the kit approach for the more casual robotocists out there. And if you're more hard-core hobbyist, stick around this chapter as well, since we will identify the needed parts and where you can get them. We will also take a very brief look in this chapter at alternate and additional parts that can be used with your robot.

Getting the Kit

As we mentioned in the previous chapter, Carnegie Mellon licensed the PPRK to Acroname, Inc. Acroname redesigned the original kit to make it easier to assemble, take apart, and ensure proper alignment of the various components. This makes it extremely easy for you to get all the parts you need to build your own robot. Acroname (check out Figure 2-1) has done all the work of getting the parts from the different suppliers and packaging them together for you. This is convenient for those of you who are just getting into the hobby of robotics and may not have a lot of spare parts lying around. On the other hand, if you happen to have a controller board, roller wheels, or death ray lasers just lying around the house, you can get the other individual components you need from various retailers. Later in this chapter we'll show you what parts you need and how to get those parts.

Figure 2-1
The front page to Acroname's web site. Not only can you get information about the PPRK, but Acroname also has information on the Sony AIBO and a new robot they've named Garcia.

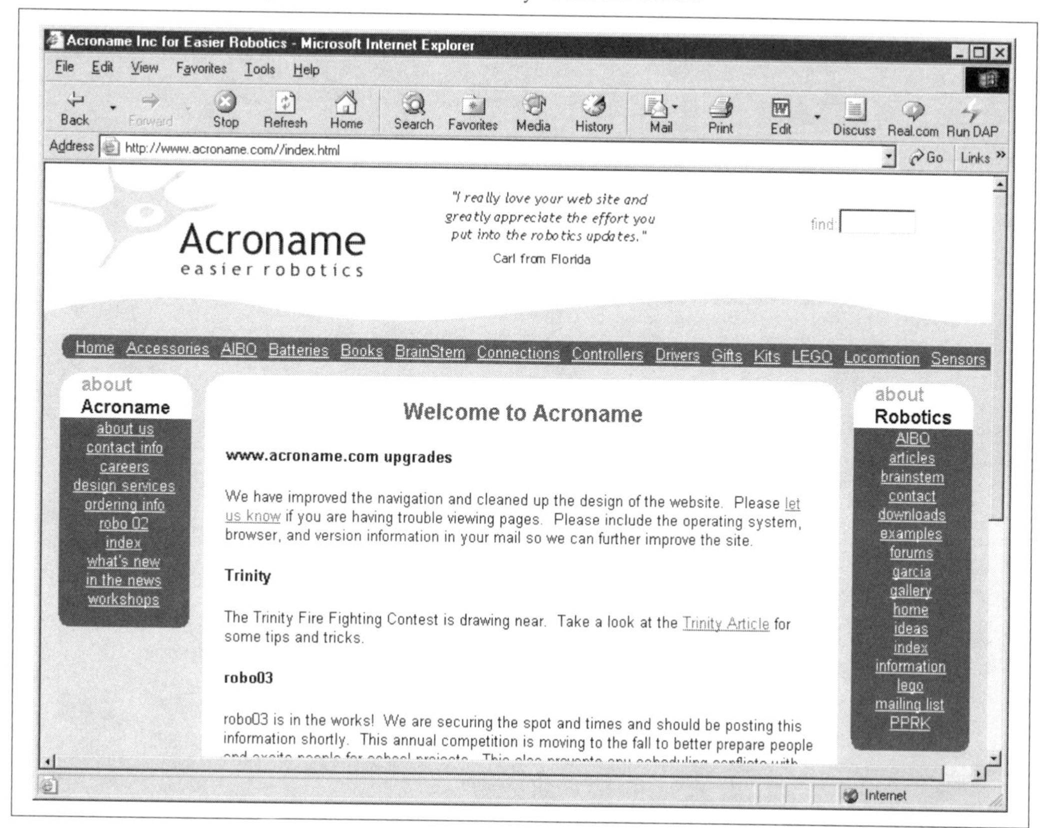

In the About Robotics box on the main Acroname page are links to the BrainStem, Downloads, and the PPRK. The BrainStem is the controller that is used in the newest versions of the PPRK. (It is a different controller from the one originally used in the PPRK.) The Downloads link will lead you to a section of the web site where you can download software to use with the PPRK. We will be exploring the downloads area later in the book when we start programming the robot. The PPRK link leads to general information about the robot. Near the top of the page is a row of links including a link labeled Kits (see Figure 2-1).

Figure 2-2
The Acroname shopping page for various robot kits. You can see the five PPRK kits in addition to a few other kits.

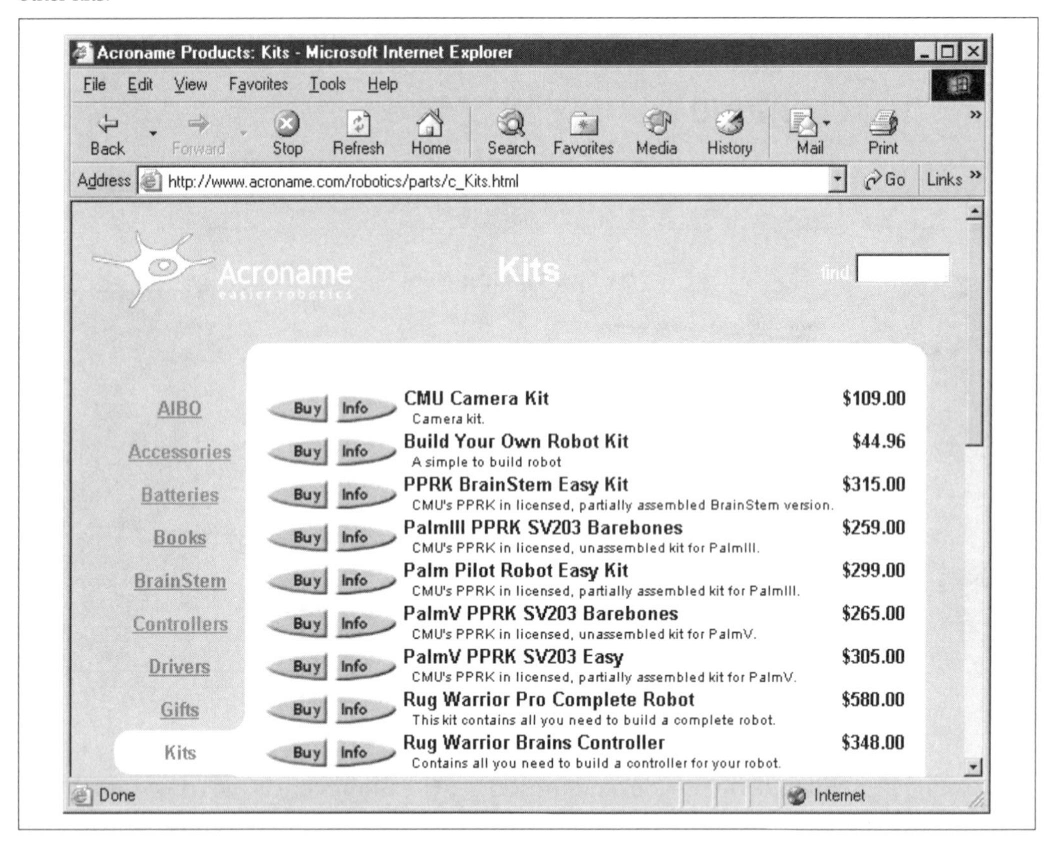

Clicking on the Kits link will take you to a page that lists all the kits that Acroname sells (see Figure 2-2). There are five different PPRKs available:

- **Palm III PPRK Barebones** The original PPRK with a connector for the Palm III. This kit includes all of the necessary parts, but no instructions. The instructions are available as a download from the Acroname web site. Constructing this kit includes modifying the *servos* (the motors) and wiring a serial cable. This kit is not for

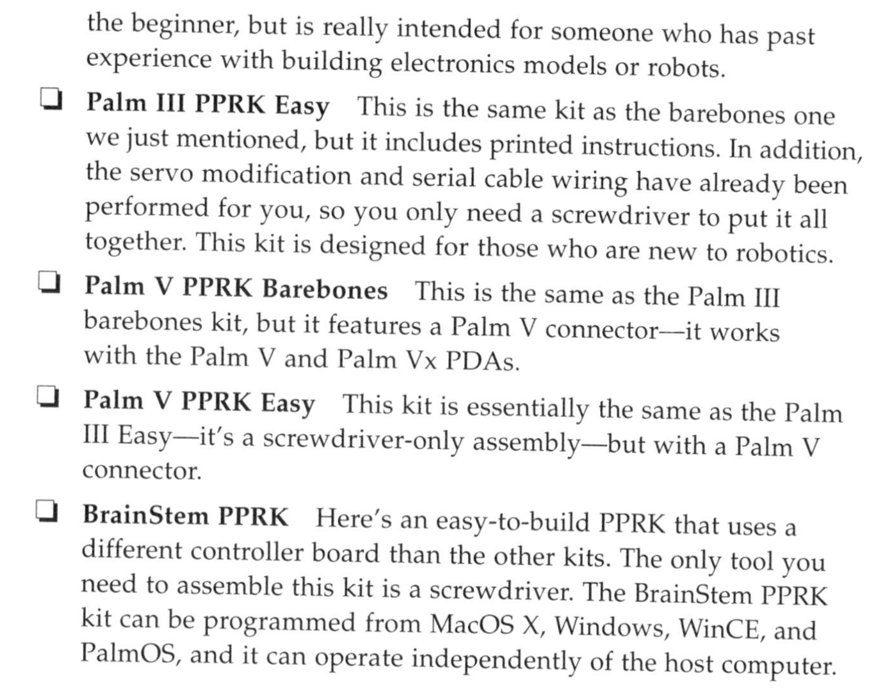

the beginner, but is really intended for someone who has past experience with building electronics models or robots.

- **Palm III PPRK Easy** This is the same kit as the barebones one we just mentioned, but it includes printed instructions. In addition, the servo modification and serial cable wiring have already been performed for you, so you only need a screwdriver to put it all together. This kit is designed for those who are new to robotics.
- **Palm V PPRK Barebones** This is the same as the Palm III barebones kit, but it features a Palm V connector—it works with the Palm V and Palm Vx PDAs.
- **Palm V PPRK Easy** This kit is essentially the same as the Palm III Easy—it's a screwdriver-only assembly—but with a Palm V connector.
- **BrainStem PPRK** Here's an easy-to-build PPRK that uses a different controller board than the other kits. The only tool you need to assemble this kit is a screwdriver. The BrainStem PPRK kit can be programmed from MacOS X, Windows, WinCE, and PalmOS, and it can operate independently of the host computer.

Want to order a kit? That part is pretty simple. You can order kits and parts online or by fax, mail, or phone. New to robotics? Then we recommend that you buy the BrainStem PPRK. It's easy to build and easy to program. Indeed, this is the kit that we will primarily be using as our example throughout this book. If you are more adventurous, you might want to consider the Palm III or Palm V PPRK. If you are really a techno geek (like Kevin), then you probably want to try one of the barebones kits—it gives you the most to do.

Getting Parts from Other Sources

If you are a robotics hobbyist, you may already have a lot of spare parts lying around the house. Having a well-stocked parts drawer is invaluable when you suddenly need a spare cross-coupling quantum magnetic capacitor at 11:00 P.M., and the local hardware stores are closed.

So open up that parts drawer and let's see what you have… pens, pencils, transistors, paper clips, magnetic capacitor, string, servos, twist ties, memory chips, glue, nickels, wire, wheels, notepad, toggle switch, heat shrink tubing, batteries, small pebbles, electrical connectors, an old 386 motherboard, an ISA sound card, and some modeling clay.

Well, that's a pretty good collection, and perhaps McGuyver could make a robot out of it. You, however, are a few microcontrollers short of an iron giant. Since you still need to collect a few parts for your PPRK, let's look at the various parts you'll need and some of the places you can get them.

If You Only Had a Brain

The original PPRK was based on the Pontech SV203 controller board (you can see a picture of it in Figure 2-3). This is the same board that you can still get in the Acroname Palm III and Palm V kits.

Figure 2-3
This is Pontech's web site showing the SV203.

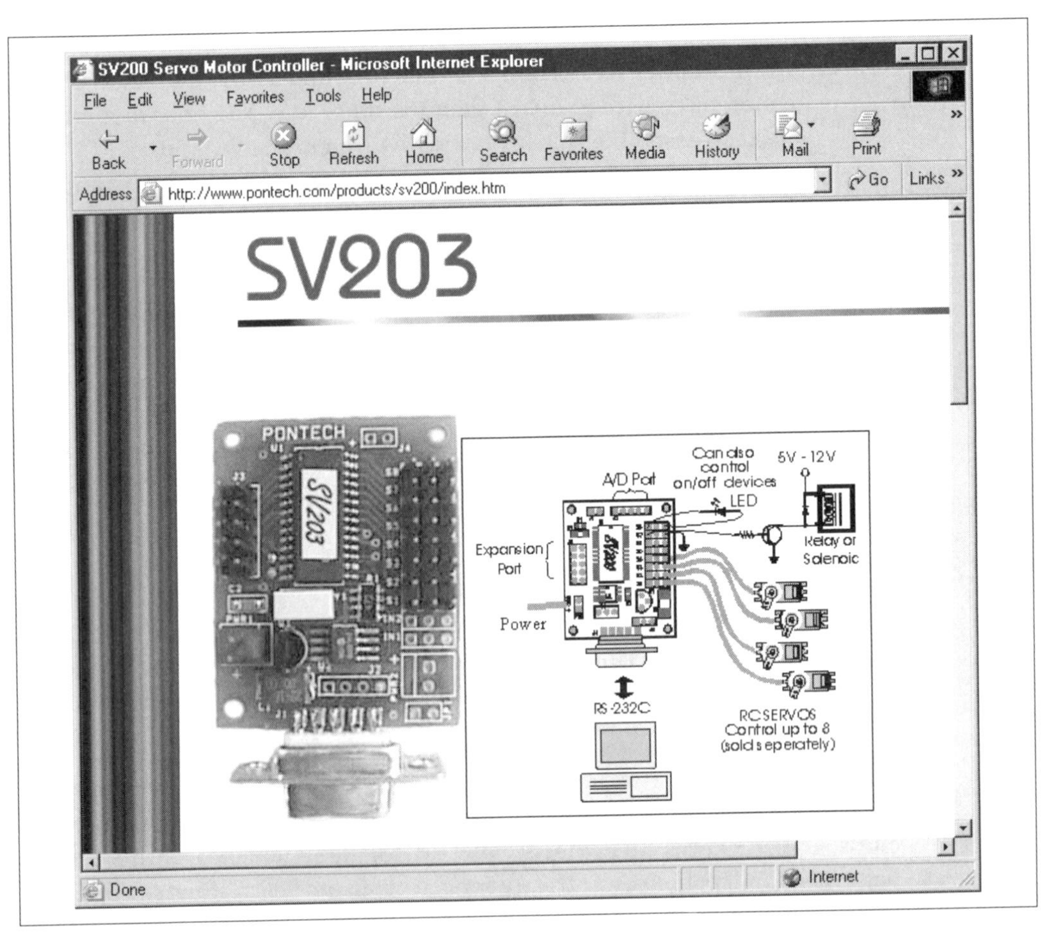

The original SV203 controller acted as an interface between the Palm and the robot. It accepted serial data (commands) from the Palm, and converted those commands into digital signals that directed the robot's servos (more commonly known to us as motors) to rotate in one direction or the other. It also read analog data from the sensor input ports, converted that to a digital value, and sent the sensor reading to the Palm. The SV203 has eight servo output ports and five sensor input ports. With the SV203, the Palm had to be connected to the controller at all times.

The newest version of the PPRK from Acroname runs on Acroname's BrainStem controller board (see Figure 2-4). This board is similar in function to the SV203. It can provide an interface between the Palm and the sensors and motors, or it can store a program downloaded from a PC or Palm, and run the robot autonomously. The BrainStem has five analog input ports, five digital input ports, and four servo controller output ports.

Figure 2-4 The BrainStem controller from Acroname

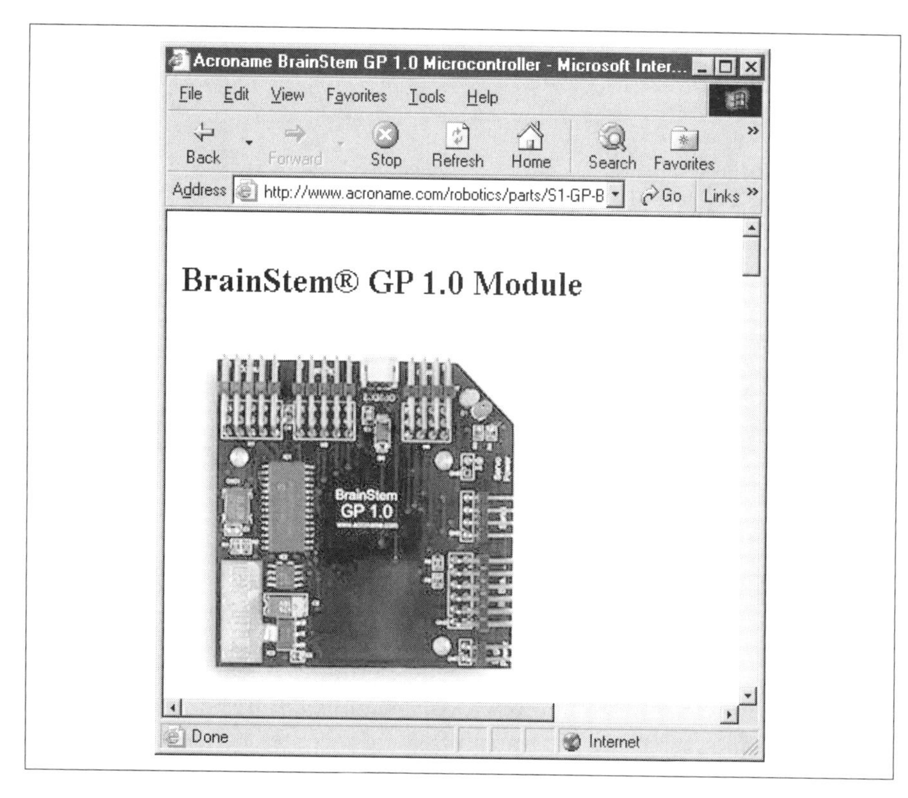

For the BrainStem controller, Acroname is the one and only source. If, instead, you want to experiment with the SV203 controller, you can get it from Acroname, from the manufacturer (that's Pontech, at www.pontech.com), or from Jameco Electronics at www.jameco.com.

See Me, Feel Me

To get information about its environment and the objects around it, the PPRK uses sensors. To be precise, the original PPRK used infrared rangers. These rangers transmit an infrared signal from one sensor; if an object reflects the signal, the return signal is received at another sensor. Using the reflected signal, the distance to a nearby object can be calculated.

The PPRK uses Sharp GP2D12 infrared rangers, seen in Figure 2-5. This ranger can detect objects that are between 10 cm and 80 cm (approximately 4 inches to 31.5 inches) away from the sensor. Sharp Microelectronics makes a number of different IR rangers. You can find detailed information about the rangers at www.sharpsma.com/sma/products/opto/optical_systems.htm.

Figure 2-5
The Sharp GP2D12 sensor transmits signals from an infrared LED and receives the return signal on the receiver. When the sensor is positioned so that the white connector is toward the bottom, the transmitter is the LED on the left and the receiver is the sensor on the right.

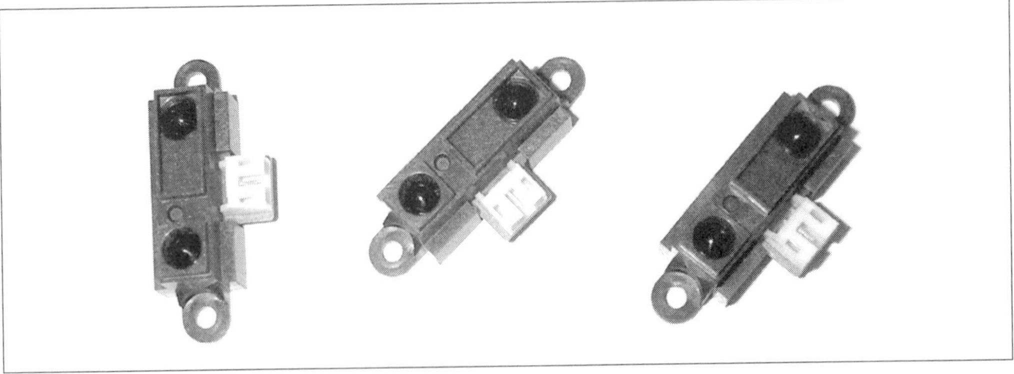

The Sharp GP2D12 ranger determines distance to objects by a technique known as triangulation. If you examine Figure 2-5, you'll see that a small distance separates the transmitter and receiver. By measuring the transmit- and receive- angles of the reflected IR signal, the ranger can calculate the distance to the object that reflected the signal.

Enough theory. Need to get your hands on some sensors? Some of the online retailers that we found carrying Sharp sensors include

- www.digikey.com/DigiHome.html
- www.arrow.com
- www.nuhorizons.com

Motor City Madhouse

As we've already mentioned, the servos are the motors that drive the wheels based on a set of commands sent by the controller. Usually, servos are built so that they rotate some set amount and then stop. This makes them well suited to moving robotic appendages that have a limited range of motion, but it makes them unsuitable for driving the wheels of a robot that need to be able to rotate continuously in one direction. To deal with this problem, you can modify a normal servo so that it can rotate continuously.

On the other hand, you can let someone else do the modification for you, and buy the servo from them (like the ones in Figure 2-6).

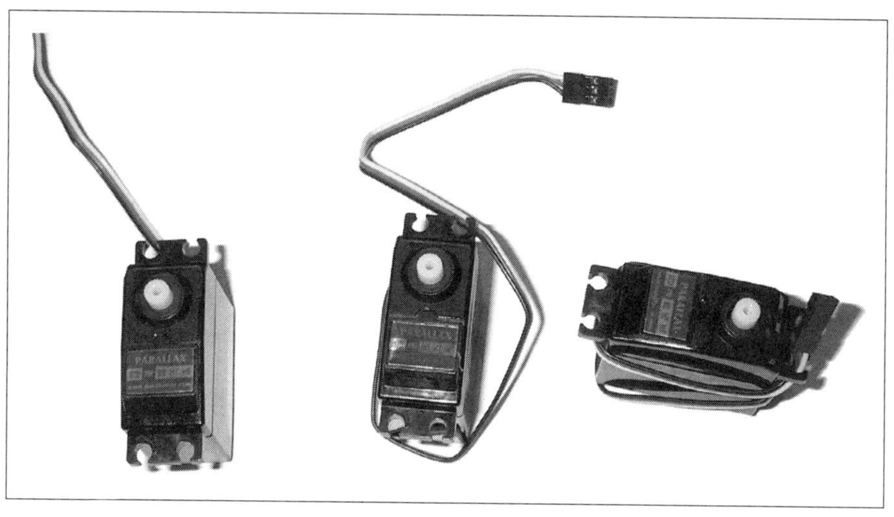

Figure 2-6 Here, you can see the servos that come with the BrainStem kit. These servos have already been modified so that they will rotate continuously.

Your best source for the servos is Acroname. Another excellent source: Mr. Robot (www.mrrobot.com). Both Acroname and Mr. Robot also have directions for modifying servos for continuous rotation.

Under My Wheels

By this time you may have noticed that the PPRK only has three wheels. (If you haven't noticed yet, congratulations! You have the observational skills of Dave.) That's not such a big deal, right? Tricycles, bicycles, and three-wheel all terrain vehicles do just fine with three wheels. The difference is that the PPRK's wheels are not aligned linearly; they are oriented in a triangular pattern such that no single wheel lines up with either of the other two wheels! Don't believe us? Check out Figure 2-7.

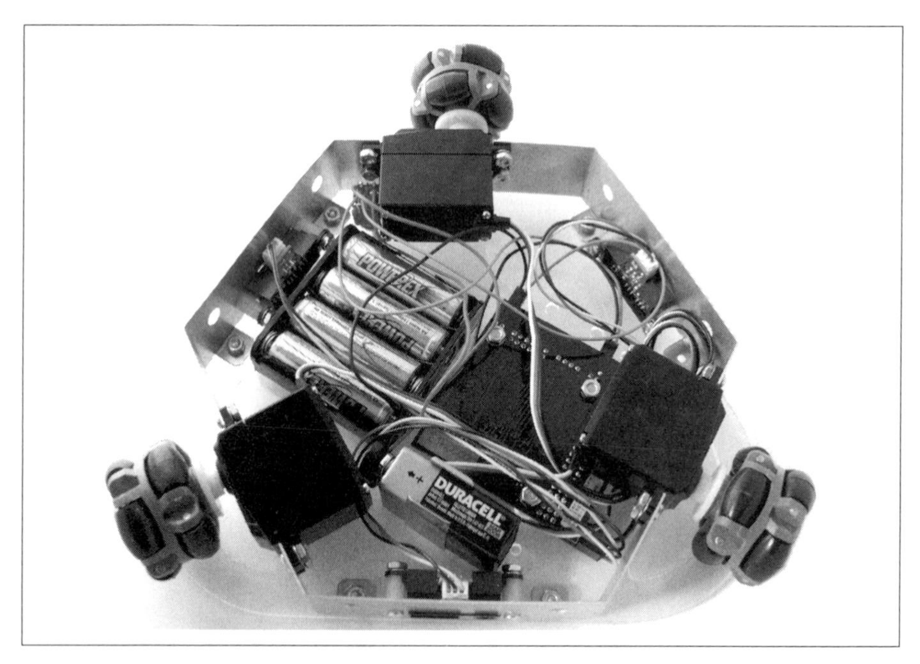

Figure 2-7 The PPRK's wheels are spaced around the frame of the robot in a triangular pattern.

This is the point where, if you have an alter ego, it is no doubt asking you, "Self, how does the robot travel in a straight line if none of the wheels are aligned?" We'd like to say the answer involves ten-dimensional string theory and quantum electrodynamics, but in truth, the answer is much more mundane.

The secret lies in the type of wheels that are used. These wheels are designed so that they can roll along two perpendicular dimensions. When the axle is rotating, the wheel rotates like a normal wheel, and the wheel moves, more or less, in the direction of rotation. However, at other times, the robot may be pushing or pulling the wheel in line with the axle. When this happens, rollers that are embedded along the radius of the wheel allow the wheel to move in line with the axle. In other words, the wheels can actually move perpendicular to their normal direction of travel. There's a close-up of this unique wheel system in Figure 2-8.

Figure 2-8 One of the omni-directional roller wheels used for the robot. When rotating around the axle, the wheel rolls like a normal wheel. But the embedded rollers also allow the wheel to pull off a clever trick: it can roll in line with the axle.

These wheels are produced by North American Roller Products (www.narp-trapo.com). Unfortunately, these folks don't sell their products directly to consumers. To purchase these wheels, you will need to go to a retailer that carries them.

In addition to Acroname, the other location we have found for these wheels is Mr. Robot (www.mrrobot.com).

Be sure you buy the right ones. The wheels you will need for the PPRK are 4 cm omni-directional wheels. Are you a monster truck fan? Cool. You'll be happy to know that both Acroname and Mr. Robot sell larger omni-directional wheels. Be sure to email us some pictures of your Monster PPRK with oversized wheels.

How Do the Wheels Work?

The ability of the PPRK to travel in a straight line relies on the physics of motion. Any force applied to an object can be resolved into component forces in any arbitrary coordinate system. Alternately, any two or more forces can be summed into a single resulting force. For example, in the normal X-Y (Cartesian) coordinate system, every force has an X component and a Y component.

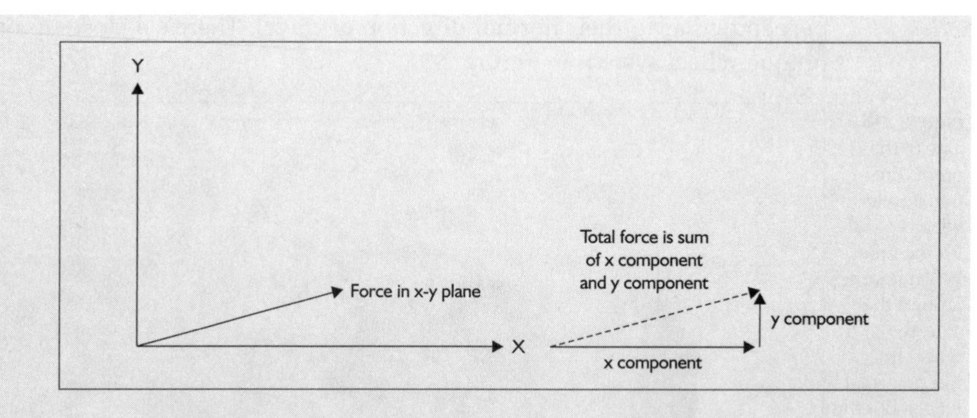

That simple principle applies to the PPRK and its wheels. Each wheel can supply a force to the robot, and the sum of the different forces resolves into a single composite force that moves the robot in any direction desired.

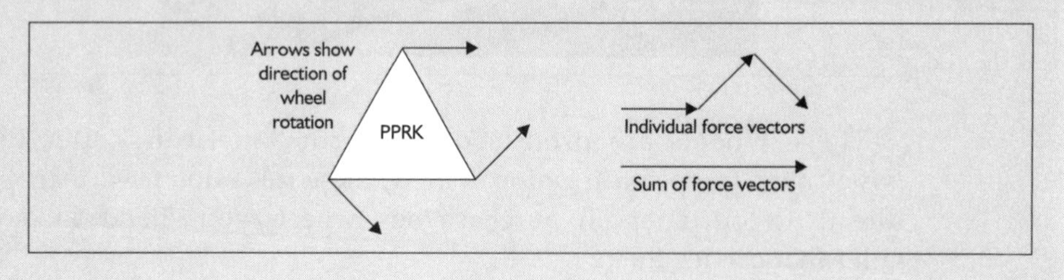

Talk to Me

Connecting the Palm III or Palm V to the SV203 controller board so that the two can talk to each other requires a special cable with a Palm connector on one end, and a DB9 serial connector on the other end. You can get a prebuilt connector for this purpose from www.syncablesolutions.com. You can also get these connectors from Acroname, but they are not prebuilt. With the connector from Acroname, you will still need to solder the wires in the Palm cable to the DB9 serial connector.

The BrainStem controller uses a different kind of connector. For the BrainStem, Acroname carries a number of different cables to connect many different devices to the controller. There are cables for Palms, Handspring Visors, PCs, and even Pocket PCs.

Odd Ones

Finally, there are a number of smaller parts that you will need when building the PPRK. Many of these parts can be obtained from local stores or Kevin's garage. If you need an essential part when you're working on your Palm Robot at 3:00 A.M., don't hesitate to give him a call. He'll drive the parts right over.

Batteries

Your robot will need a power source. We initially toyed around with incorporating a small fusion power source, but had trouble containing free neutrons in the heavy water reservoir. As a result, we recommend that you stick with regular batteries.

The BrainStem kit uses four AA NiMH rechargeable batteries (although any AA battery will work) to power the servos, and a 9-volt battery to power the controller board and IR rangers. If you are using the SV203 kit, you'll need four alkaline AA batteries. Obviously, you can find these batteries just about anywhere. You'll also need to get a battery holder for the batteries. Any local hobby store should carry these.

Acrylic Platform

The frame, controller board, and batteries are mounted on a single piece of clear acrylic. These can be obtained from any hardware store and cut to size. The original PPRK used a circular platform, cut to six inches in diameter. You can purchase a precut acrylic disc at www.mcmaster.com. (Don't be intimidated by the sheer number of parts sold there; just search for part number 8581K26, and it'll pop up in a heartbeat.) If you're taking the kit approach, you'll find that the PPRK kits from Acroname use a slightly larger platform that's cut into a triangular shape with rounded corners. If you are going to duplicate the platform provided by Acroname, you'll want to get a piece of acrylic (between 8 and 12 inches) that you can cut to the size you prefer.

Toggle Switch

The robot can be turned on and off using a toggle switch. These can be found at various electronics or hardware stores. You can find them online at Jameco Electronics (www.jameco.com: look for Jameco part 105814). You can also substitute similar switches from other sources, like Radio Shack, which sells more varieties of toggle switches than you could ever possibly need.

Connector Housing

Connecting the various sensors, servos, and battery packs to the controller uses wires with simple connectors so that no soldering or special wiring is needed. Simply plug the wires into the connector housing, connect the housing to the appropriate pins, and the connection is complete.

All of the connectors mentioned here can be found at www.mouser.com. All you need to do is simply search for the part number. Another source is www.acroname.com, where you can find a grab bag of various connectors for your robot.

For the PPRK that uses the Pontech controller, you will need five Molex connectors, part number 50-57-9002.

For the BrainStem controller, you'll need two pieces of Molex 50-57-9002 and six pieces of Molex 50-57-9003.

For all the Molex connectors, you'll need crimp terminals. These crimp around the wire and slide into the housing. This mates the wire to the housing. For the Molex 50-57-900x connectors, you can use Molex 70058 or 71851 crimp terminals.

To connect to the sensor, you'll need three pieces of Molex 87369-0300. You will need Molex 50212 crimp terminals with this connector housing.

Aluminum Frame and Miscellaneous Hardware

The aluminum frame connects to the acrylic deck, and is used to support the servos and IR rangers. The miscellaneous hardware consists of nuts, bolts, washers, and nylon spacers to assemble the PPRK. Not surprisingly, the best place to get the frame and the hardware is Acroname. The Acroname frame comes in three pieces and is already cut and drilled to the correct size; the hardware is already selected to fit the holes drilled into the frame.

Miscellaneous Parts

This stuff is available at any hardware store:

- ❑ **Double stick foam tape** This is for attaching the battery holder to the acrylic disk.
- ❑ **24 AWG wire** You'll use this stuff for connecting various components. We recommend getting at least three different colors for identification purposes.

Carry On

At some point, you will want to go beyond the servos and IR rangers that come with the basic PPRK. Later in this book, in fact, we will do just that. You can find a number of other sensors, including a flame detector, a compass, and a line detector, all of which you can integrate into your Palm Robot.

A great web site for learning how to interface your PPRK to other sensors and devices is Robert's Gadgets and Gizmos: www.bpesolutions.com/gadgets.ws/gproject4.html.

A few of the alternate sensors mentioned at Robert's Gadgets and Gizmos include a light sensor, a temperature sensor, and a touch sensor. Both Robert's and the Acroname web sites also include information on how to interface your robot with other output devices (as in Figure 2-9). In addition to operating the servos, for instance, the controller board from Acroname can be fitted with a voice module. Robert's Gadgets and Gizmos shows you how to use the controller to turn light emitting diodes (LEDs) on and off. This barely scratches the surface, of course, of what you can find at Robert's Gadgets and Gizmos, so be sure to spend some time exploring that site (see Figure 2-10).

Figure 2-9 The Acroname web site showing the line detector sensor

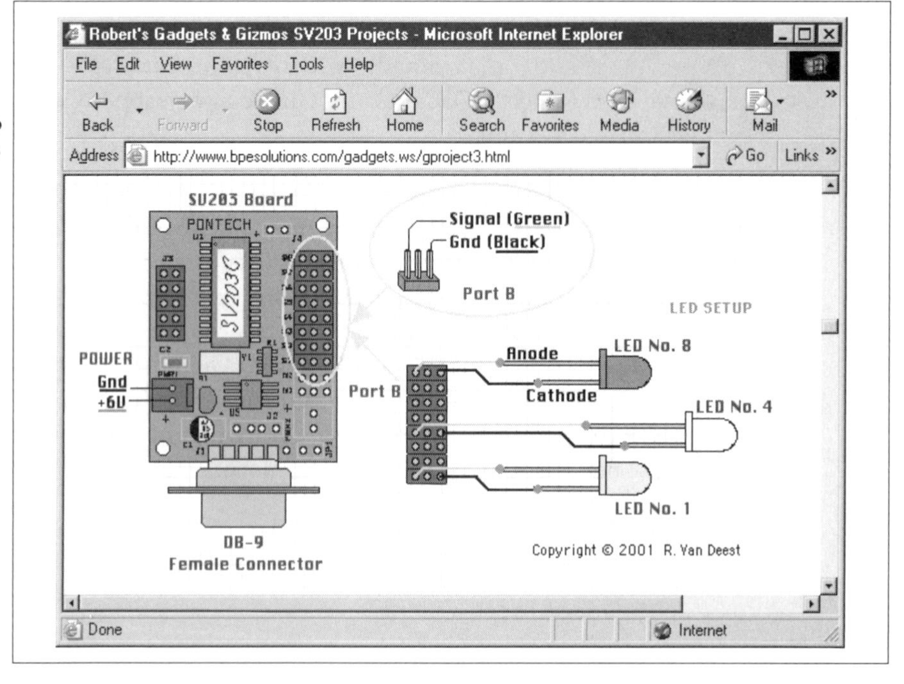

Figure 2-10 Robert's Gadgets and Gizmos shows how to connect LEDs to the output ports of the SV203 controller.

Where in the World...

In this chapter, we've listed numerous online resources for finding parts for your robot. So that you don't need to go skimming through the chapter to find every reference, we'll list them all again here.

Remember that because of the dynamic nature of the Web, some of these sites may not still be accessible using the exact URLs we've listed here—if you can't find the page you're looking for, search for it, since the information may have moved. Also, keep in mind that we haven't personally tried all of the retailers listed here in this book, so purchase stuff with caution (always good advice when dealing with web-based stores). We did use several criteria when listing these retailers, though. Each retailer is either the manufacturer of a part, an authorized dealer for the manufacturer, or a retailer that we know has a stable business.

All parts needed for PPRK:

www.acroname.com

Controller boards

www.pontech.com/cgi/pricelist.pl

www.jameco.com

Cables

www.syncablesolutions.com/

Connectors

www.mouser.com

Sensors

www.digikey.com/DigiHome.html

www.arrow.com/

www.nuhorizons.com/

Wheels

www.mrrobot.com

Miscellaneous

www.allelectronics.com/

www.jameco.com/

www.bpesolutions.com/gadgets.ws/gproject4.html

Finally, you may want to check out *Robot Builder's Sourcebook : Over 2,500 Sources for Robot Parts*, by Gordon McComb (ISBN: 0071406859) for a general purpose listing of sources.

Chapter 3

Building the Robot

By now, you've gotten a kit, or the parts you need, and are ready to start building the robot. If you have one of the barebones kits from Acroname, or you collected the parts yourself, you'll want to start at the beginning of this chapter—"Building the PPRK with the Pontech Controller."

If you have one of the easy kits, you can skip ahead to the section "Building the Frame." If you have the BrainStem kit from Acroname, you'll want to skip ahead to the section "Building the PPRK with the BrainStem Controller."

Finally, if your kit has many different little colored plastic pieces that look like Legos, you have the wrong book. May we recommend that you check out *Robot Invasion*, which Dave wrote; it has three different projects that build fun robots out of Legos.

Building the PPRK with the Pontech Controller

Before attempting to put anything together, you should examine everything in the kit and compare it to the shipping list to ensure that you have everything you are supposed to have. Here is a list of all the parts that should be in the kit:

- ❑ Three omni-directional wheels
- ❑ Three IR sensors
- ❑ Four rechargeable batteries and a battery charger
- ❑ One Pontech SV203 controller
- ❑ One DB9 male connector
- ❑ One Palm connector
- ❑ Three servos
- ❑ Three aluminum frame pieces
- ❑ One toggle switch
- ❑ One acrylic deck
- ❑ Three white connector housings
- ❑ Five black connector housings
- ❑ One red connector housing
- ❑ Nine wires—three each red, yellow, and black
- ❑ Nuts, washers, and screws

If you are building the barebones kit, you will also need a soldering iron, solder, heat shrink tubing, epoxy, a standard screwdriver, and a Phillips screwdriver.

Did you take the easy way, and buy the easy kit? If so, grab a standard screwdriver and a Phillips screwdriver, and jump ahead in this chapter to the section "Building the Frame." If you have the barebones kit, you have a little more work to do first.

The Batteries

The kit contains rechargeable batteries to power the robot. Since you can't do much with uncharged batteries, start by putting the batteries into the charger and plugging the charger into an electrical socket. This will ensure that the batteries are ready for you when the robot is complete.

The Safety Geek Says

This next step involves a process called *soldering*. If you've built electronic circuits before, you're probably already familiar with how to do this and how to do it safely. If you've never soldered a connection before—STOP! Find someone who knows what they are doing and ask them to give you a quick lesson in how to solder connections. Here are some further tips:

- Solder contains lead, and lead can be poisonous. Take care to wash your hands after using lead solder.
- Don't inhale the fumes from the soldering process.
- The soldering iron, hot enough to melt solder, is hot enough to give you a nasty burn. Treat the iron with care. Unplug it when you are finished.
- The connection you just soldered is as hot as the iron. Give it a chance to cool before you touch it.

Wiring the Battery Pack

Your battery pack should have a black wire and a red wire extending from it. Start by soldering the red wire to the middle post of the switch.

Your kit should have another red wire with a crimp on one end, and exposed wire at the other end. Solder the exposed wire to one of the outer terminals of the switch—it doesn't matter which.

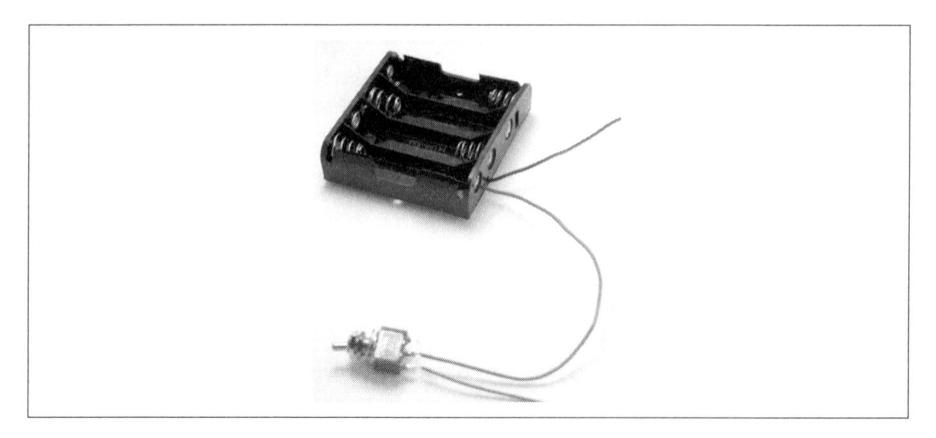

The black wire from the battery pack, like the red wire from the previous step, has a crimp on the end. These crimps are inserted into the red connector housing. The connector housing fits onto the power pins on the SV203 controller, and it only fits in a particular orientation. With the housing oriented to match the board, insert the red crimp into the side of the housing that will fit into the + pin of the controller. Insert the black crimp into the other side.

Modifying the Servos

Next, we need to modify the servos. Start by removing the white wheel from each servo.

Carefully remove the screws from the bottom of the servo housing.

Carefully remove the lid from the servo.

Remove the topmost gear and the output gear. The topmost gear is the center gear. The ouput gear is the gear with the white shaft that extends out of the servo lid. Underneath the output gear is a small rectangular brass pin. In a moment you will use this pin to adjust the servo.

Attach the servo to the controller board at pin S1. The black wire in the connector plugs into the pin closest to the edge of the board. Put batteries into the battery pack and plug it into the controller.

The servo may start rotating at this point. If it does not, toggle the power switch so that the servo starts rotating. Now turn the brass pin until the servo stops rotating. The servo is now centered. Detach the servo from the controller and turn the power switch off.

Find the large gear that you removed earlier. It has a tab on it that stops the servo from rotating fully. Clip this tab off the gear.

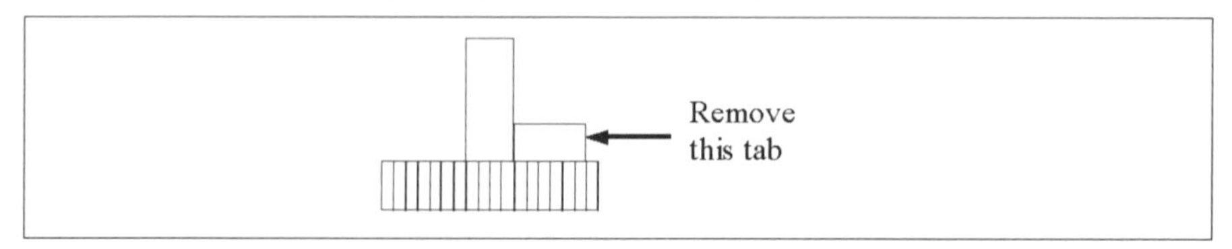

The tab prevents the gear from rotating continuously, so you need to ensure that it is completely removed. We used a diagonal-cutting plier to cut most of the tab and an exacto knife to trim the remainder. Also, use caution when cutting the tab, so that you don't cut yourself.

If you bought the kit from Acroname, the kit came with six ball bearing assemblies. The assemblies are small ring-shaped pieces that hold the ball bearings. Place one over the brass pin. Place another on the horn of the output gear.

Reassemble the gears and put the lid back onto the servo. Carefully reattach the servo to the controller and turn the power switch on to test that the servo is still centered. If it is not, you will need to disassemble and recenter the servo. Reinsert the long screws so that the lid is reattached to the servo body. Put this servo aside and perform the same procedure for each of the other servos.

Assembling the Wheels

Take the white plastic wheel you removed from each servo and the three omni-directional wheels. You are going to glue one white wheel to the axle hole of each omni-directional wheel. Orient the white wheel so that it can be screwed onto the servo in its original orientation. (The ridged hole in the white wheel faces away from the omni-directional wheel and fits over the servo horn.) Use sandpaper to roughen the two surfaces that will be glued together. Place a small amount of epoxy on the white wheel and put the two wheels together. Before the glue sets, center the white wheel in the axle hole; you can check the alignment by looking through the other end of the axle hole.

Repeat the previous step for each wheel.

Attaching the Palm Connector to the Deck

There are two 1/4-inch holes drilled along one side of the deck. (There are smaller holes drilled along the other two sides.) Alternately, your acrylic deck might have a 1/4-inch hole and a slot along one side of the deck. You want the side with the 1/4-inch holes (or the hole and the slot). Peel back the paper from the acrylic, and position the Palm connector so that it is centered between the two holes.

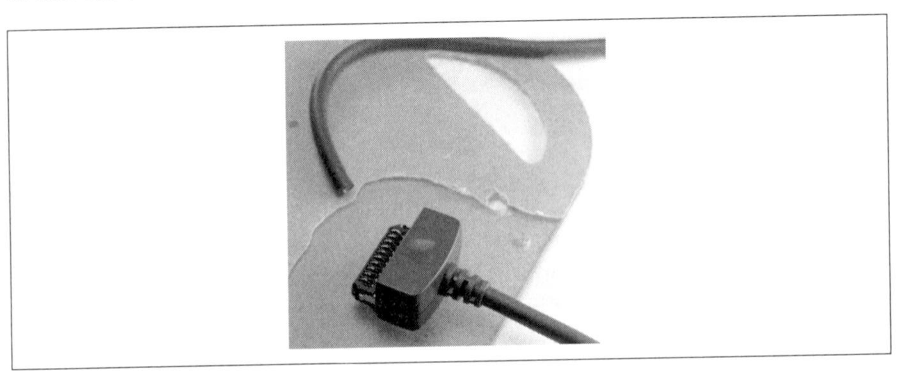

Mark the position of the connector. Now slightly roughen the surface of the acrylic with some sandpaper. Using double-sided foam tape, attach the Palm connector to the deck. When the connector is secure, thread the Palm cable through the hole or slot to the left.

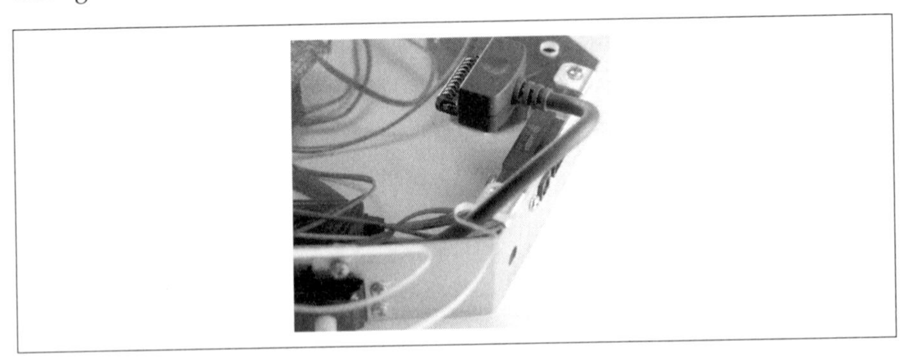

Cut the Palm cable to 14 inches. Remove approximately 2 inches of the outer insulation from the cable. Remove all the string, shielding, and paper so that you have clear access to all six wires in the cable. Based on the following table, select the three wires you need, and clip the other three wires back.

		Acroname Palm III Cable Kit	Acroname Palm V Cable Kit	Palm III Hotsync Cable
DB9 Pin	2	brown	green	brown
	3	yellow	purple	black
	5	blue	black	red

Strip enough insulation from the three wires so that you can solder them to the DB9 male connector. If you are using heat shrink tubing for insulation, slip a small piece onto each wire. Solder the wires to the DB9 connector as shown in the table. Shrink the tubing around the soldered connection.

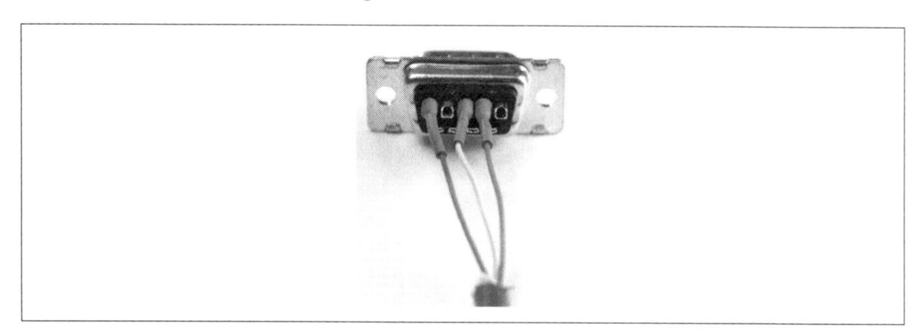

Building the Frame

Start by finding the IR rangers, and the parts to connect the ranger to the aluminum frame. For each ranger you will need two 5/8-inch screws, two nylon washers, two nylon spacers, two lock washers, and two nuts. These parts will be in bag one of the kit. Attach one ranger to each aluminum frame piece as shown next. Start by threading both screws through the frame, then adding the nylon washers and spacers. Place the sensor onto the screws and fasten it with the lock washers and nuts. Be careful not to overtighten the nuts.

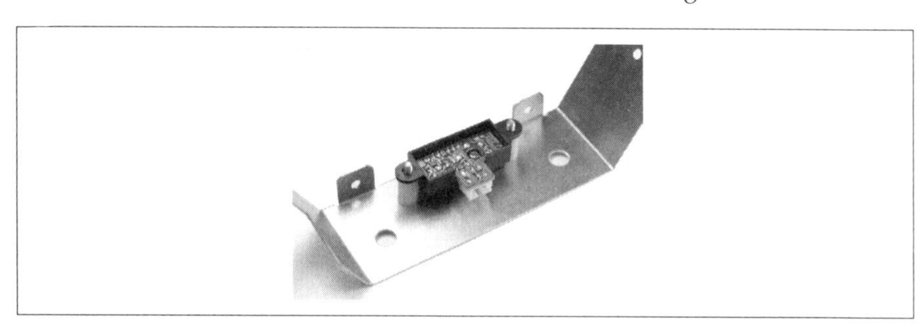

Now you will attach the three frame pieces together to make a single frame. Notice in the next illustration that each piece has a long arm and a short arm. When the pieces are oriented correctly, the long arm on one piece will line up with the short arm on an adjacent piece. Align the holes in each arm with the long arm to the inside of the frame, and connect the pieces using a screw, washer, and a nut from bag two of the kit. The washer and nut will be to the inside of the frame. Tighten the nuts just enough to hold everything together; we will tighten them completely in a later step.

The next step is to attach the servos to the frame. Each of the three servos will be attached to the frame through the rectangular hole that is on each side of the frame where the frame pieces connect. You can see that the hole is offset from the center of the side; when the servo is placed correctly in the hole, the white knob extending from the servo will be centered in the frame. Place a servo into the frame and attach it to the frame using the 1/4-inch screws, the nuts, and the washers from bag two of the kit. Note that the servo flanges are mounted inside the frame.

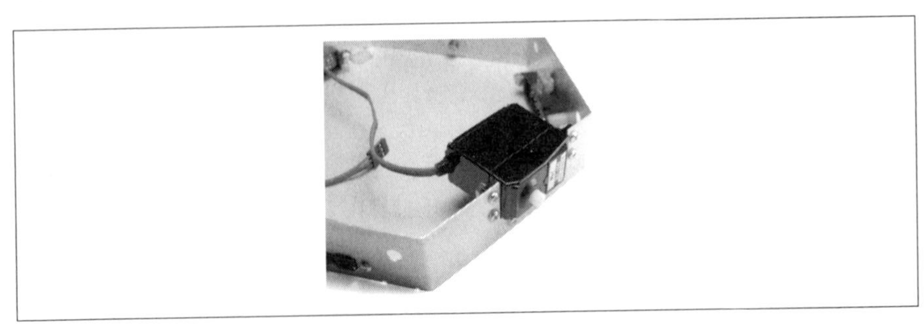

When each servo is securely attached to the frame, tighten all the nuts on the frame and the servos. You will want to perform this on a flat surface so that you ensure the frame remains flat when all the screws and nuts are tightened.

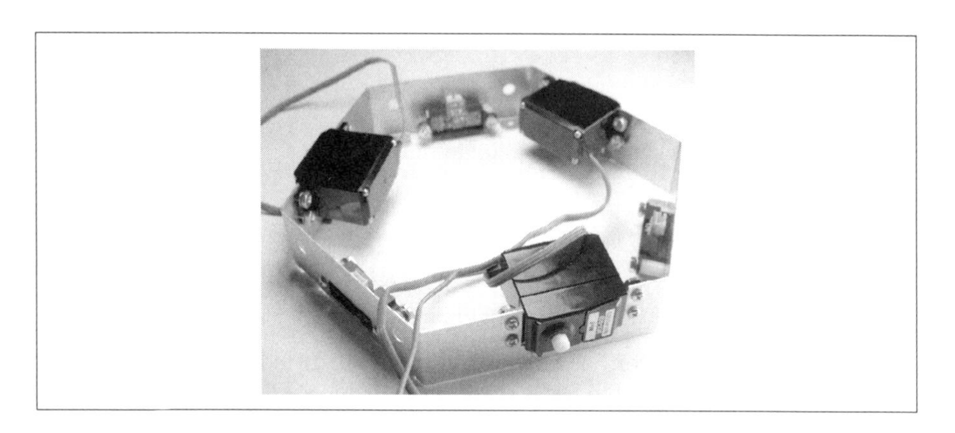

The Controller and the Deck

Start by peeling the paper off the acrylic. The acrylic deck has holes predrilled into it. These holes will be used to attach the controller and a few other parts. Since the acrylic deck is designed to support several different controllers, you will not use all the holes.

The Robot Geek Says

The Acrylic Deck

If you are building the robot from scratch, and you bought an acrylic deck that was not predrilled, you will need to drill your own holes in the deck. Here is what you need:

- Three countersunk holes for the BrainStem controller
- One 1/4-inch hole for the switch
- One hole (or slot) for the controller cable
- Holes for attaching the deck to the frame (if needed)

As you read through the chapter, make note of the general location of these holes, and use that as a guide for positioning and drilling the holes for your robot.

From bag two, find the three nylon spacers that have a threaded end, three nylon screws, and three nylon washers. Insert a screw through one of the countersunk holes in the deck, add a nylon washer, and then screw the nylon spacer onto the screw. Repeat for the other two spacers.

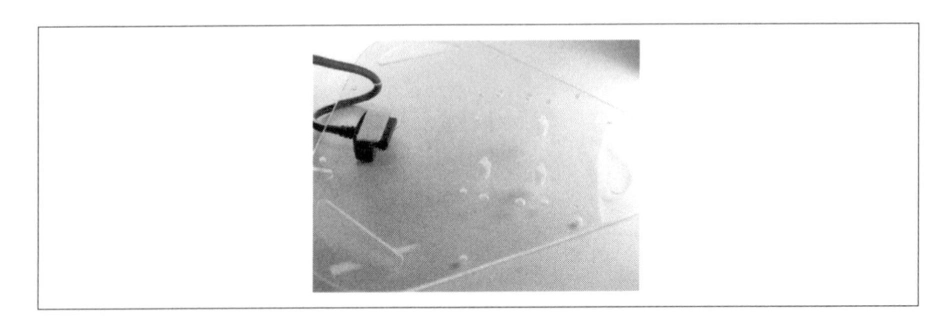

Each spacer has a threaded post. Position the Pontech controller on top of the posts, so that each post goes through a mounting hole on the controller board. Attach the board to the spacers using a lock washer and hex nut.

Attaching the Deck to the Frame

There are six metal tabs spaced around the top of the frame. These tabs should line up with holes drilled into the acrylic deck. Line up the deck with the tabs, and attach them with screws, washers, and nuts.

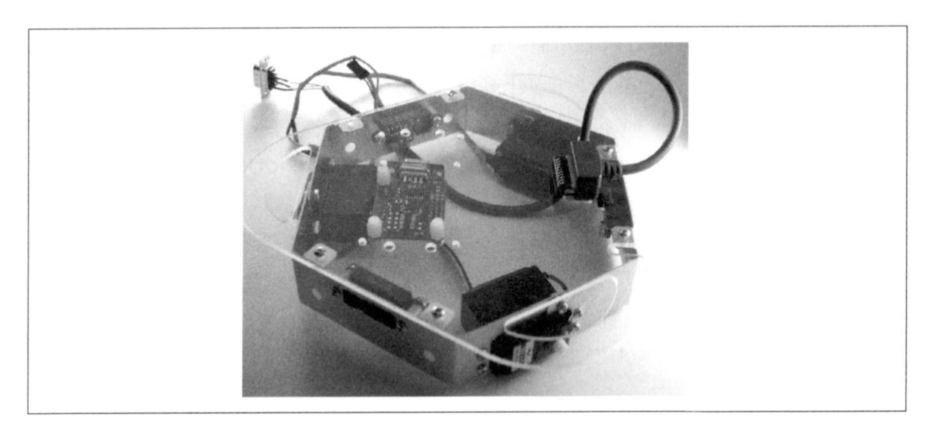

Completing the Wiring

Find the wires that came in bag three of the kit. You will find three red wires, three yellow wires, and three black wires. The primary reason for having three colors is to keep track of which wire is which when you are inserting the wires into the Molex housing. These wires already have crimps on each end. The short crimps fit into the white housing that connects to the IR sensors; the long crimps fit into the black housing that connects to the controller board. Looking at the following illustration, you will see that each crimp has a small tab that extends up from the crimp. This tab will lock into a slot on the housing and prevent the wire from being easily removed from the housing.

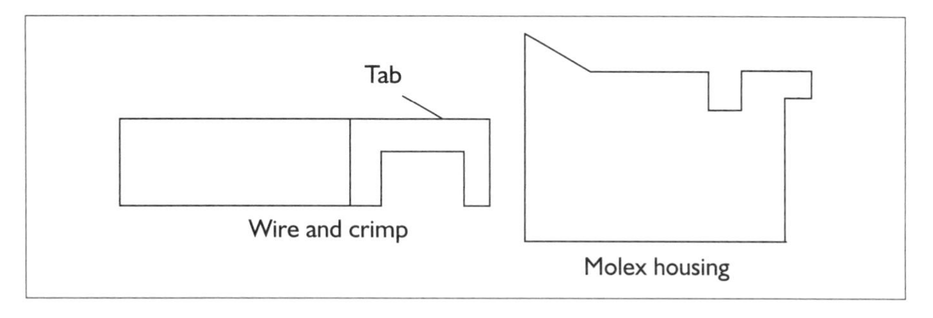

Start with the short crimps and the white Molex housing. Insert the wires into the housing as shown next. With the housing oriented as shown in the illustration, the wires from left to right are red, black, and yellow. The middle wire on this connector is the ground wire.

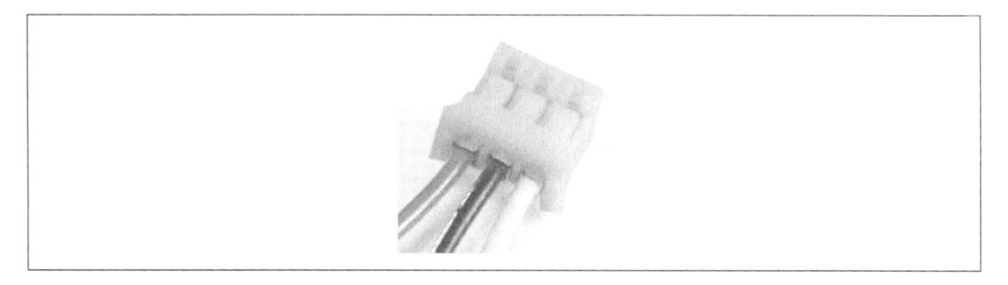

Now take each pair of black and red wires, and insert them into a black Molex housing. This will leave all the yellow wires loose; we will get to them soon.

Connect one white Molex housing to each sensor. There is a tab on the wire housing that fits into a slot on the sensor housing.

Now you will connect the other end of each wiring assembly to the SV203. If you look at the component side of the SV203 controller, with the DB9 connector toward the bottom, you will see a block of pins on the right side of the board labeled S1 through S8. The wire assembly from Sensor 1 should be plugged into S8, with the black wire closest to the edge of the board. Likewise, Sensor 2 plugs into S7, and Sensor 3 plugs into S6.

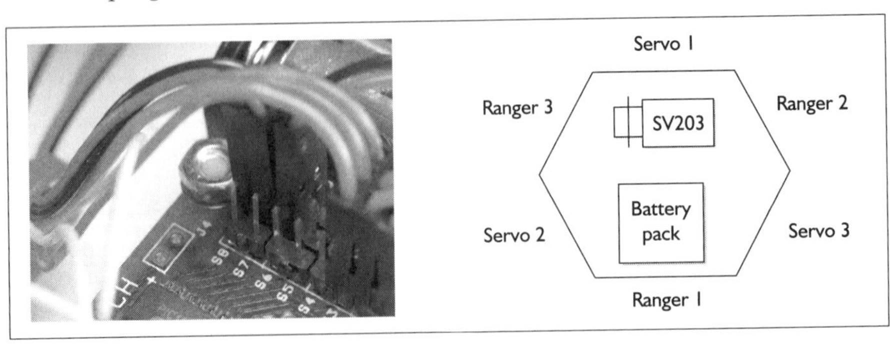

Take the yellow wires from IR Rangers 1 and 2, and insert them into a Molex housing. Take the final yellow wire from Ranger 3, and plug it into the remaining housing. On the left side of the controller board is a block of pins labeled J3. Attach the first housing to the topmost set of pins, with the wire from sensor 1 closest to the edge of the board. Plug the final controller into the next set of pins, with wire three plugging into the pin closest to the edge of the board.

Now you will perform a similar operation with the servos. The wire assembly from Servo 1 plugs into S1; Servo 2 plugs into S2; Servo 3 plugs into S3. In each case, the black wire should plug into the pin closest to the edge of the board.

The next step is to mount is the switch. The switch will have two washers and two nuts. Remove the topmost nut and washer. Spin the bottom nut so that it is about two-thirds of the way down the threads on the switch as shown in the following illustration. The washer on top of this nut should have a small tab that fits into a slot in the threads. Now insert the switch into the deck so that the toggle extends above the deck. Use the other washer and nut to fasten the switch to the deck. You may need to adjust the vertical position of the bottom nut to ensure that the switch fits well.

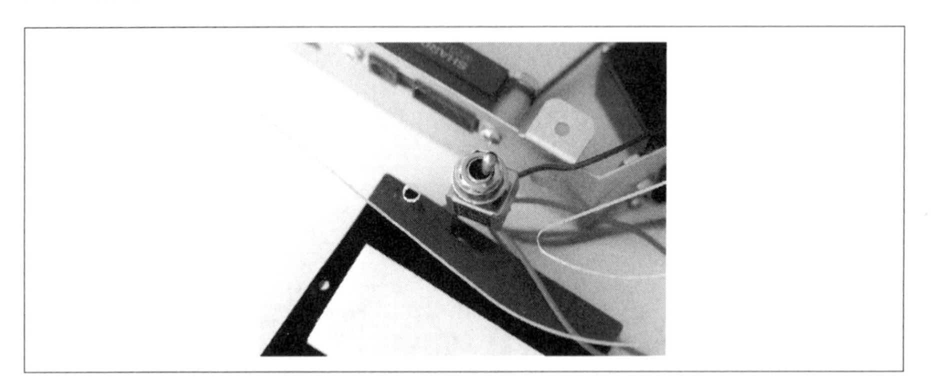

Finally, position the battery pack into the frame. When you have it positioned so that it fits correctly, secure it in place with the foam tape. Take the battery wire assembly, route it to the controller board, and plug it into the power pins on the left side of the SV203. The black wire will plug into the pin closer to the center of the board. It should fit only when it is positioned correctly.

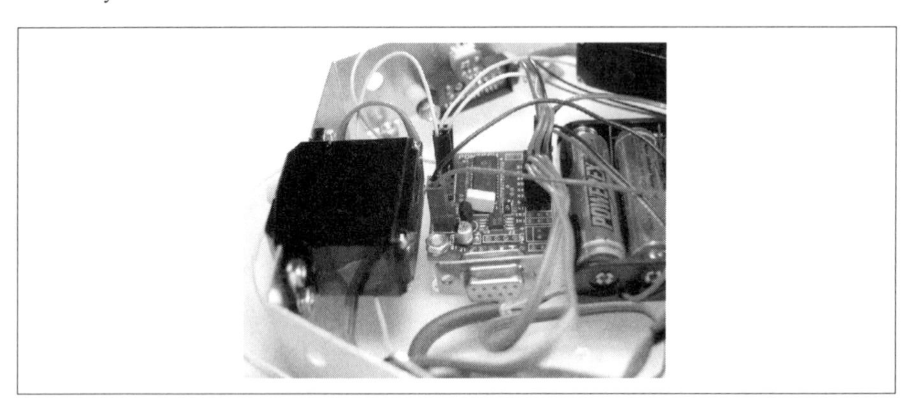

Attaching the Palm Connector and the Wheels

Take the DB9 end of the Palm connector and plug it into the DB9 connector on the SV203 controller. Use a cable tie to bundle the Palm cable and the other wires underneath the deck. The cable tie loops through two holes in the acrylic. Leave a small piece of the cable tie protruding above the deck. Attach another cable tie to the other side of the deck. The two cable ties help to keep the Palm in place.

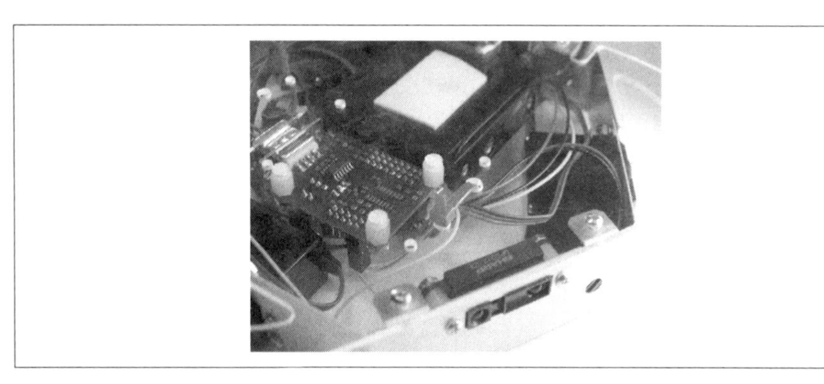

With the deck attached to the frame, you're ready for the wheels. Each wheel has a white disk on one side. This disk will fit over the white horn extending from the servo. The hole in the disk has small teeth that fit into similar teeth on the horn. With the teeth lined up properly, the wheel will slide onto the horn with just a little pressure. Use the small black screws from bag three to attach the wheels. Be careful to get the screw lined up properly, and don't strip the screw hole.

The robot is now complete. If your batteries are charged up, insert them into the battery holder. If the wheels move or twitch when the batteries are inserted, the switch is in the On position, so move it to the other position.

What's Next?

If you are using the Palm III or Palm V kits, you are ready to start playing with your robot. You can skip Chapter 4 and move straight to Chapter 5. In

Chapter 5, we'll show you where to get software for your robot, and how to use the software with your robot. If you want to do your own programming, go to Chapter 6 to see how to use the Palm Robot Programmer, or Chapter 7 to see how to program the Robot using C/C++, BASIC, or Java (Chapter 8).

If you are using the PPRK with a Palm VII, you still have some work to do. Proceed to Chapter 4 to see how to do this.

Building the PPRK with the BrainStem Controller

Like the PPRK easy kit, all you need to build the BrainStem kit is some patience and a couple of screwdrivers—one standard and one Phillips. Before attempting to put anything together, you should examine everything in the kit and compare it to the shipping list to ensure that you have everything you are supposed to have. Here is a list of all the parts that should be in the kit:

- ❑ Battery charger, four rechargeable AA batteries, one 9-volt battery
- ❑ Three poly roller wheels
- ❑ Three servos
- ❑ One acrylic chassis plate
- ❑ One BrainStem controller
- ❑ One bag of parts containing the aluminum frame, the sensors, and the various screws, washers, and nuts to connect the sensors to the frame
- ❑ One bag of parts containing the screws, nuts, and washers for holding everything else together
- ❑ One bag of parts containing the wires, crimps, and battery holders

Additionally, you should have purchased an interface cable of some kind. Since the BrainStem can interface to many different platforms, the interface cable is sold separately from the kit.

The Batteries

The BrainStem kit contains rechargeable batteries to power the robot. Since you can't do much with uncharged batteries, start by putting the batteries into the charger and plugging the charger into an electrical socket. This will ensure that the batteries are ready for you when the robot is complete.

Building the Frame

Start by finding the IR rangers, and the parts to connect the ranger to the aluminum frame. For each ranger you will need two 5/8-inch screws, two nylon washers, two nylon spacers, two lock washers, and two nuts. These parts will be in bag one of the kit. Attach one ranger to each aluminum frame piece. Start by inserting both screws through the frame, then adding the washers and spacers. Place the sensor onto the screws and fasten it with the lock washers and nuts.

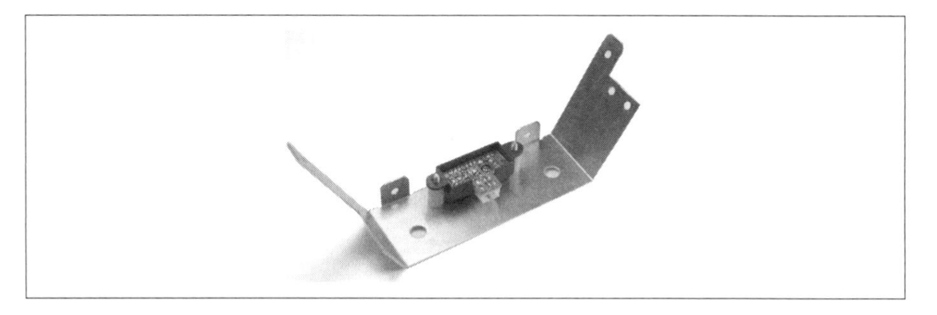

Now you will attach the three frame pieces together to make a single frame. Notice that each piece has a long arm and a short arm. When the pieces are oriented correctly, the long arm on one piece will line up with the short arm on an adjacent piece. Align the holes in each arm, and connect the pieces using a screw, washer, and a nut from bag two of the kit. Tighten the nuts just enough to hold everything together; we will tighten them completely in a later step.

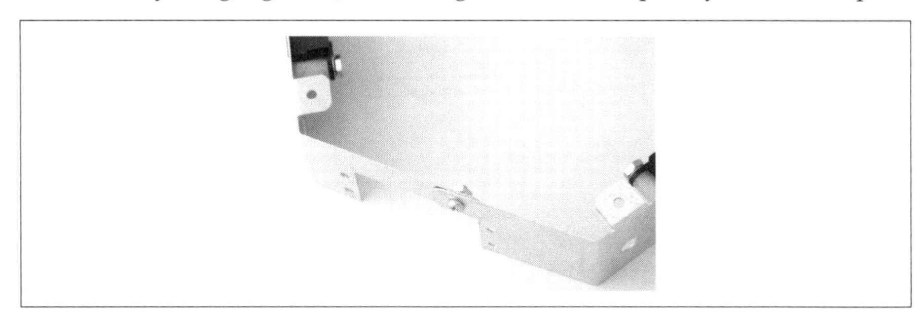

The next step is to attach the servos to the frame. Each of the three servos will be attached to the frame through the rectangular hole that is on each side of the frame where the frame pieces connect. You can see that the hole is offset from the center of the side; when the servo is placed correctly in the hole, the white knob extending from the servo will be centered in the frame. Place a servo into the frame and attach it to the frame using screws, nuts, and washers from bag two of the kit. You will need four of each to attach the servo.

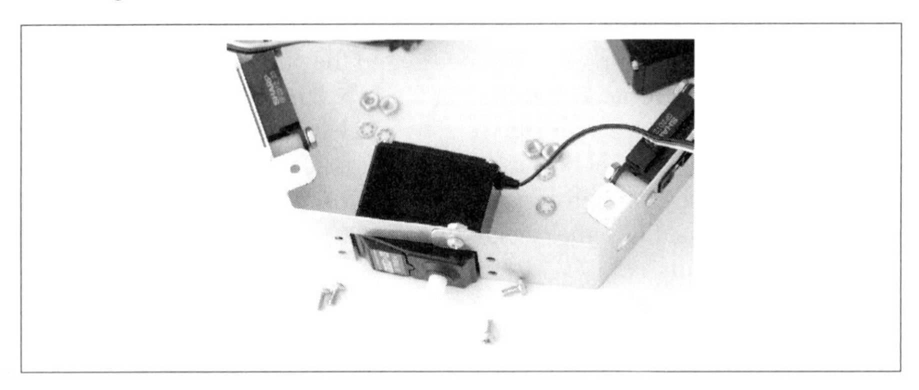

When each servo is securely attached to the frame, tighten all the nuts on the frame and the servos. You will want to perform this on a flat surface so that you ensure the frame remains flat when all the screws and nuts are tightened.

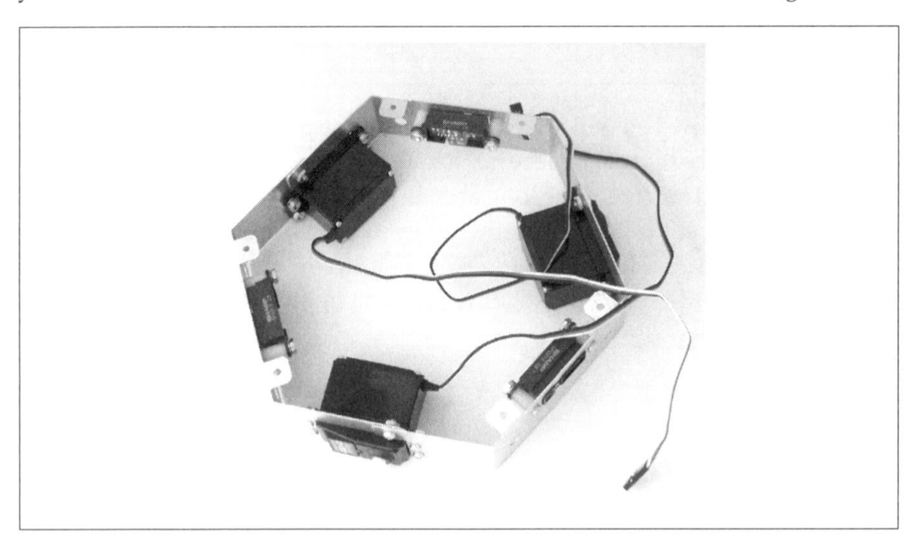

The Controller and the Deck

Start by peeling the paper off the acrylic. The acrylic deck has holes predrilled into it. These holes will be used to attach the controller and a few other parts. Since the acrylic is designed to support several different controllers, you will not use all the holes.

When we built our BrainStem PPRK, we found that the hole for the switch was not in the best location for us. We subsequently moved the switch to a new location. Because it's better to drill new holes before you attach any components to the deck, this would be the best time to drill a new hole for the switch. Drill a 1/4-inch hole as shown in the illustration.

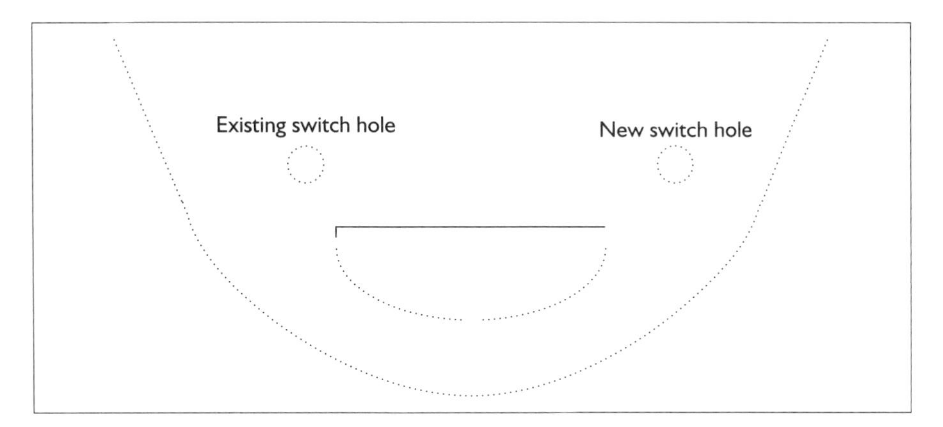

From bag two, find the three nylon spacers that have a threaded end, three nylon screws, and three nylon washers. Insert a screw through one of the countersunk holes in the deck, add a nylon washer, and then screw the nylon spacer onto the screw. Repeat for the other two spacers.

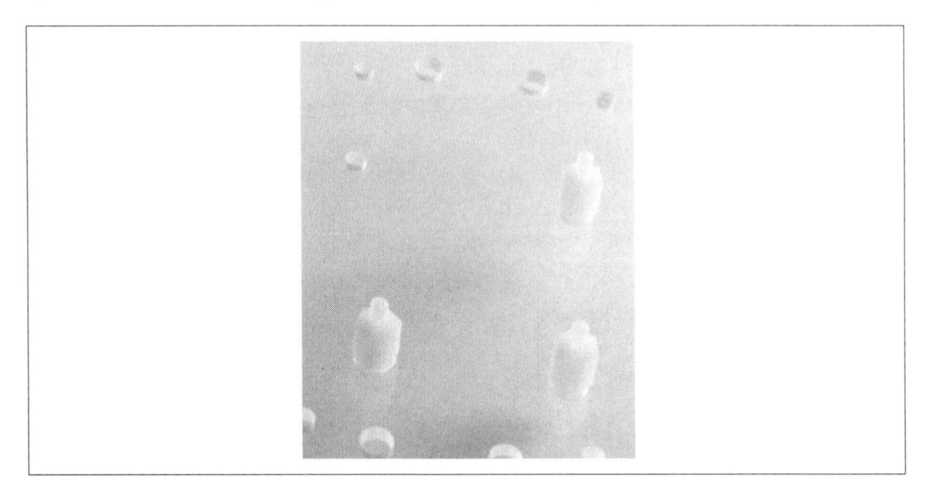

Each spacer has a threaded post. Position the BrainStem controller on top of the posts, so that each post goes through a mounting hole on the controller board. If you're confident that everything is correct, attach the board to the spacers using a lock washer and hex nut.

If you're not so sure, don't attach the board until you connect the sensors and servos and test the connections.

Completing the Wiring

Find the wires that came in bag three of the kit. You will find three red wires, three yellow wires, and three black wires. The primary reason for having three colors is to keep track of which wire is which when you are inserting the wires into the Molex housing. These wires already have crimps on each end. The short crimps fit into the white housing that connects to the IR sensors; the long crimps fit into the black housing that connects to the controller board. Looking at each crimp, you will see that it has a small tab that extends up from the crimp. This tab will lock into a slot on the housing and prevent the wire from being easily removed from the housing.

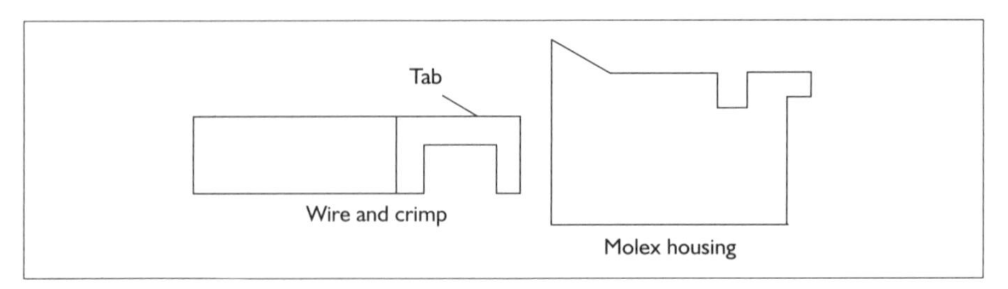

Start with the short crimps and the white Molex housing. Insert the wire into the housing. The middle wire on this connector is the ground wire.

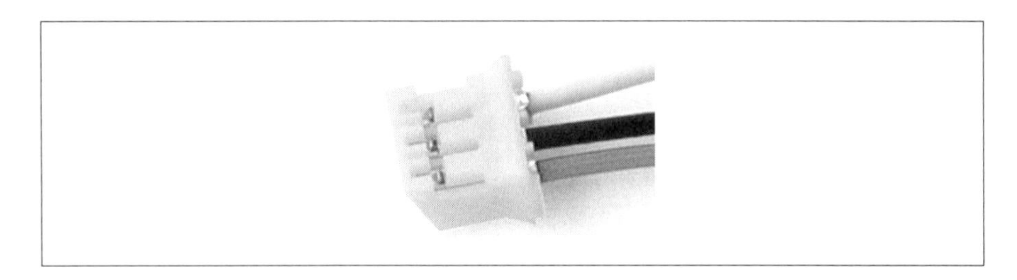

Now, you may want to wrap or braid the wires so that they are easier to manage. Acroname recommends braiding the wires. You could also wrap them with electrical tape or heat shrink tubing.

Now you will need to insert the long crimps into the black Molex housing. You must ensure that the wires are inserted in the correct order. If you examine the housing, you will see a small arrowhead imprinted on the housing. The wire that was the middle wire for the white connector should be inserted into the slot that lines up with the arrowhead. The other two wires should be inserted as shown next. The red wire goes into the center slot, and the yellow wire into the other outside slot.

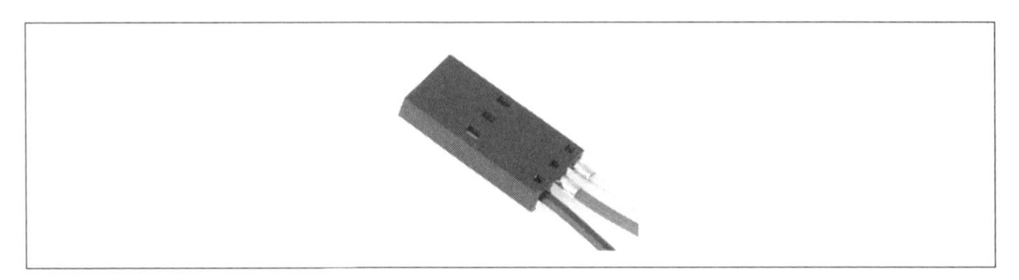

Repeat until you have three wire assemblies.

The Robot Geek Says

Wiring

You must ensure that the connector housings are wired correctly. If you attach the wires incorrectly, your robot will certainly not work; at worst, you might damage the controller board.

- In the sensor housing, the ground wire must be the center wire.
- In the controller board housing, the ground wire will be the wire that lines up with the small arrowhead imprinted on the housing.

- On the controller board, the ground wire is the pin closest to the board.

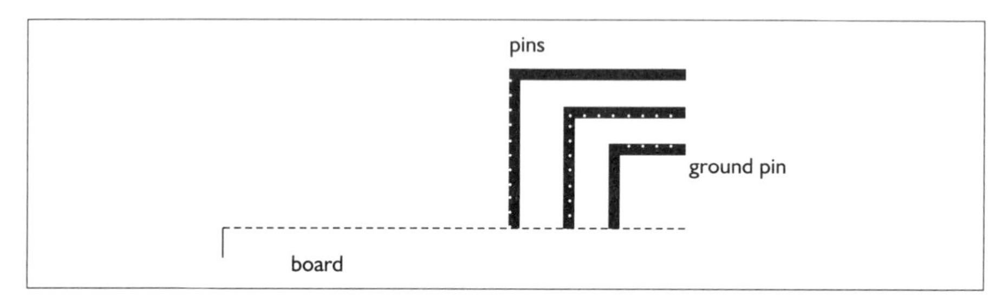

The connectors from the sensors will plug into the analog ports on the controller board. These ports are on the top left side of the board and are labeled H1. There are five ports, numbered 0 to 4; port 0 is the leftmost port, and you should be able to see the number 0 printed on the controller board under the pins for port 0.

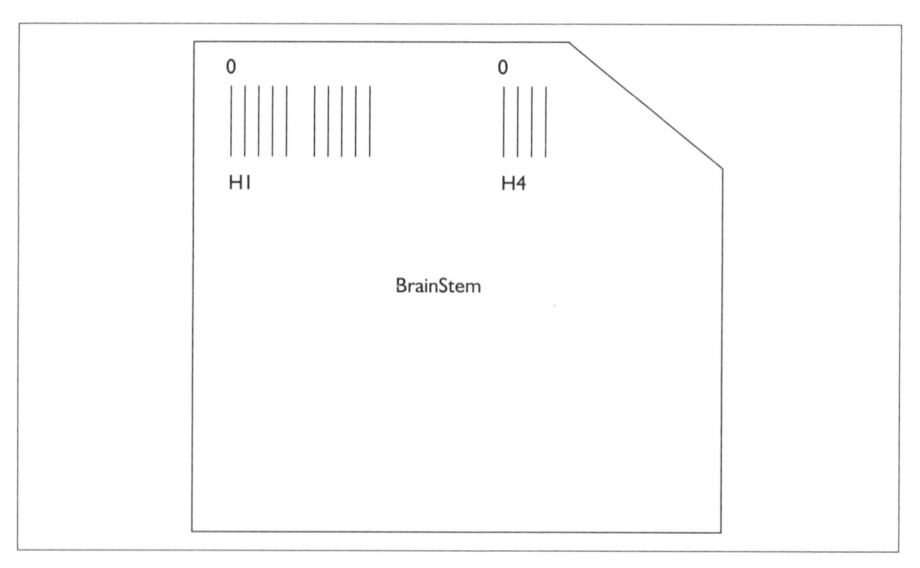

The Robot Geek Says

Port Numbering

If you are not a programmer, you might wonder why the ports are numbered from 0 to 4 and not 1 to 5. That's because of the way digital computers address and use memory locations. To put it simply, each memory location has an address. When a block of memory is used, the first location has a particular address, the next location is at address+1, the next location is address+2, etc. Thus, the first location is address+0. And that's why computer geeks who go to football games can be heard to yell, "We're number zero!"

Using the three wire assemblies, connect Sensor 1 to Port 0, Sensor 2 to Port 1, and (can you guess what's coming?) Sensor 3 to Port 2. Note that the white connector can only be inserted one way: there is a small tab on the housing that fits into a slot on the sensor connector. On the controller board, there are no tabs and slots to help you get lined up. So, ensure the ground wire (the black wire, if you're using color coding), is closest to the board, and attach the connector to the posts.

Now, you'll attach the servos to the controller. The digital ports for the servos are the four sets of pins at the top right of the controller board labeled H4. Again, the ground pin is the pin closest to the board. Attach Servo 1 to Port 0, Servo 2 to Port 1, Servo 3 to Port 2.

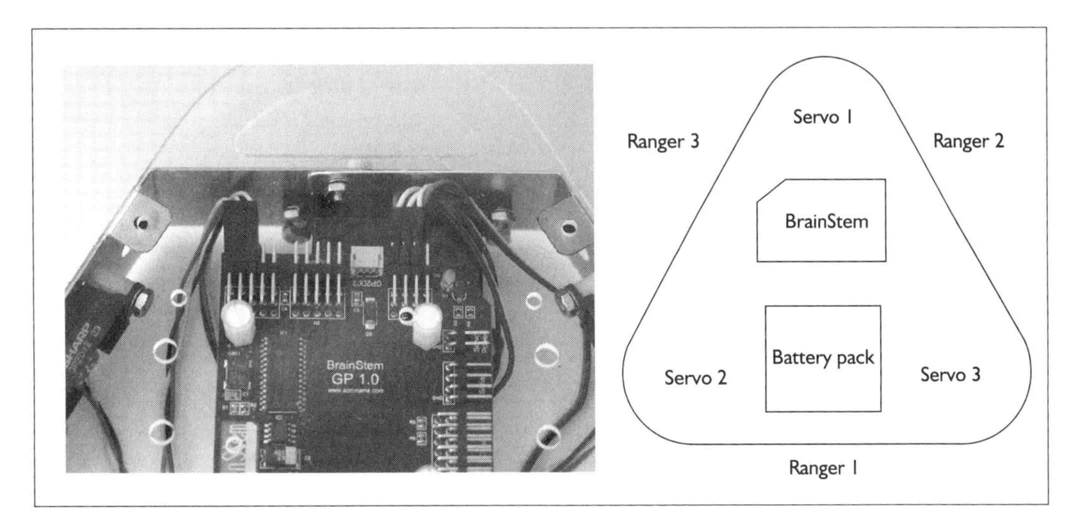

The last connections to be made are from the AA battery pack and the 9-volt connector. The battery connectors should be attached to the controller board as shown in the following illustration. The black wire of each connector is the ground wire; it should be attached to the ground pins on the controller board.

The ground pins are labeled GND on the board (although this might be partially obscured by the nylon spacer).

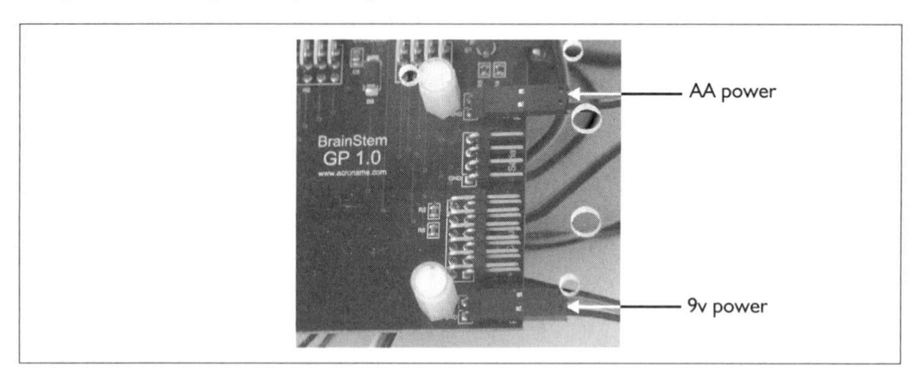

The next part to mount is the switch. On our kit, the switch was actually cramped by the servo next to it, so we later moved the switch. If you already drilled a new hole as explained earlier in this chapter, you can mount the switch in the new location. The switch will have two washers and two nuts. Remove the topmost nut and washer. Spin the bottom nut so that it is about two-thirds of the way down the threads on the switch. The washer on top of this nut should have a small tab that fits into a slot in the threads.

Now insert the switch into the deck so that the toggle extends above the deck. Use the other washer and nut to fasten the switch to the deck. You may need to adjust the vertical position of the bottom nut to ensure that the switch fits well.

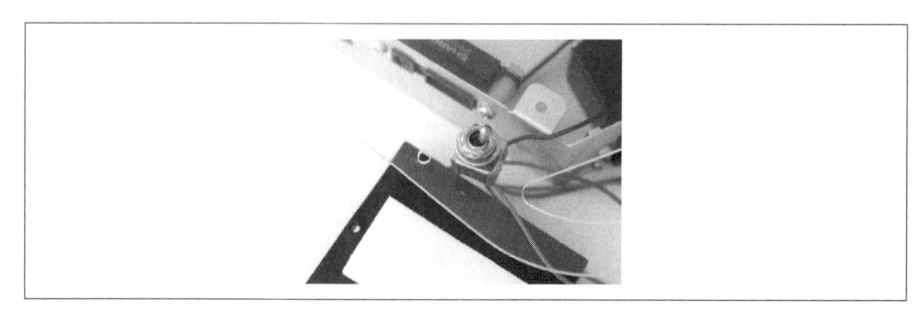

Position the battery holder in the robot. There should be room for the battery holder between the BrainStem controller and sensor 1. Attach the battery holder using double-sided foam tape.

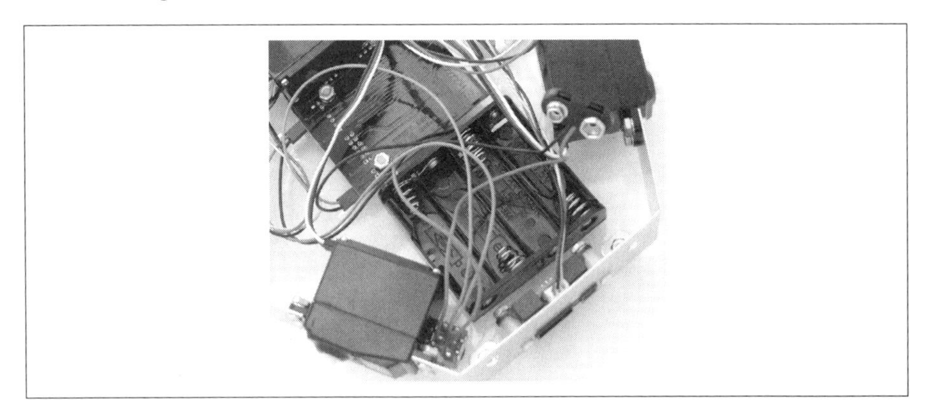

Position the 9-volt battery clip in the robot. You will want to insert the battery into the clip when you position it. The battery clip will fit opposite the power connections of the BrainStem controller. When the battery is positioned so that it fits, use double-sided foam tape to attach it.

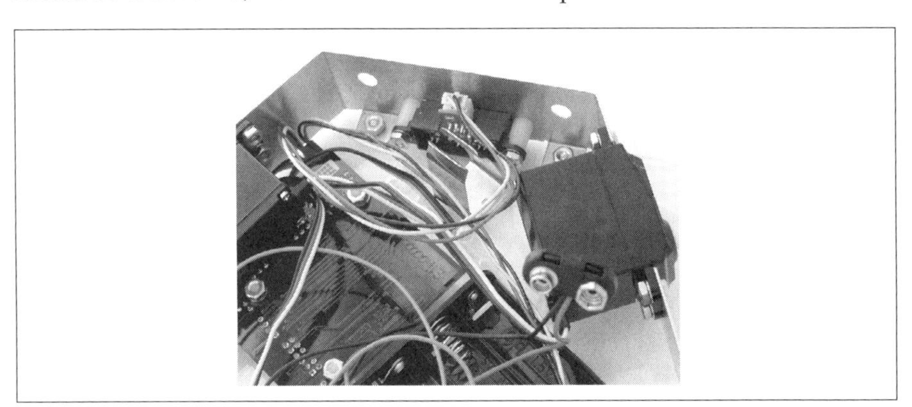

Attaching the Deck to the Frame

Before attaching the deck to the frame, you may want to test your connections. To do this you will need the Config program from Acroname. We show you how to download and install this program in Chapter 5. So, go download, install, and run Config. Use it to verify that the servos and sensors are wired correctly and are working. If they are not, check your wiring and power connections and fix any problems you find. Once everything is working, attach the BrainStem to the deck, if needed. There are six metal tabs spaced around the top of the frame. These tabs should line up with holes drilled into the acrylic deck. Line up the deck with the tabs, and attach them with the remaining screws, washers, and nuts.

Attaching the Wheels

With the deck attached to the frame, you're ready for the wheels. Each wheel has a white disk on one side. This disk will fit over the white horn extending from the servo. The hole in the disk has small teeth that fit into similar teeth on the horn. With the teeth lined up properly, the wheel will slide onto the horn with just a little pressure. Use the small black screws from bag three to attach the wheels. Be careful to get the screw lined up properly and don't strip the screw hole.

Attaching the Palm

The connector from the Palm device (or any other device you might be using) uses another Molex connector to attach to the BrainStem controller. The cable runs through a slot in the deck, between the servos and the deck, to the BrainStem controller. The cable plugs into the serial connectors on the right side of the board. Ensure that you plug the housing into the pins so that the ground wire (the wire with the red stripe) plugs into the ground pin. Most of the cables from Acroname come preassembled. One exception is the cable for the Handspring Visor. Instructions on how to assemble this connector are in Chapter 4.

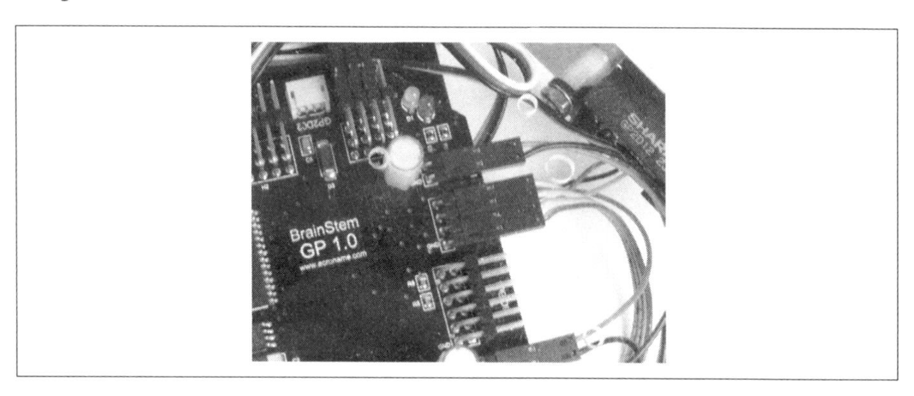

What's Next?

Your robot may not be complete, depending on which model you have and what you are using as the brains of the robot.

If you are using the Palm III or Palm V kits, or the BrainStem kit with anything other than a Handspring Visor, you are ready to start playing with your robot. You can skip Chapter 4 and move straight to Chapter 5. In Chapter 5, we'll show you where to get software for your robot, and how to use the software with your robot. If you want to do your own programming, go to Chapter 6 to see how to use the Palm Robot Programmer, or Chapters 5, 7, or 8 to see how to program the Robot using C/C++, BASIC, or Java.

If you are using the BrainStem kit and a Handspring Visor, or the PPRK with a Palm VII, you still have some work to do. Proceed to Chapter 4 to see how to do this, then Chapter 5 to get software for your robot; Chapter 6 to learn how to use the Palm Robot Programmer; or Chapters 5, 7, or 8 to learn how to program the Robot using C/C++, BASIC, or Java.

Chapter 4

Using Palm VIIs and Handspring Visors

Has this ever happened to you? You buy some cool gadget, take it home, open the box, and are then confronted by those three dreaded words: "Batteries not included." What a bummer.

If you were planning to use a Handspring Visor with your PPRK (Palm Pilot Robot Kit) or BrainStem robot, you probably feel just that way. Even though the Visor runs the Palm OS, it is easier to use a Palm III or a Palm V with your PPRK or BrainStem than to use the Handspring Visor. The Pontech SV203 won't work with the Visor at all, unless you modify the controller board. And with the BrainStem, every other style of connector cable comes preassembled, except for the Visor cable.

If you have a Palm VII, the situation is a little better. Because the Palm VII uses the same connector as the Palm III, you can use any of the PPRK Palm III kits or the BrainStem kit with the Palm III connector with your Palm VII. *Almost*. When using the Palm VII with the Pontech controller, you may need to do a software mod to the controller to get the Palm VII and the controller to communicate correctly.

In this chapter, we'll start by looking at how to construct the BrainStem connector for the Handspring Visor. Next we'll look at how to modify the Pontech SV203 to work with a Handspring Visor and how to build a connector for the SV203 and Visor. Finally, we'll look at how to use a Palm VII with the PPRK and Pontech controller.

Using a Handspring Visor with the BrainStem

The BrainStem kit from Acroname is designed to be connected to any one of many different computing devices. For this reason, the connector is sold separately from the kit. Acroname sells connectors for connecting the BrainStem to a Mac PC, a Windows PC, a Linux PC, a Palm III, a Palm V, and a Handspring Visor. For most of those devices, the cable comes already built and ready to attach the BrainStem to the device.

Unfortunately, if you have a Handspring Visor, you've probably already discovered that the cable is not already built and that you've got some soldering to do.

Parts List

These parts should come in the cable kit from Acroname. If, however, you are trying to roll your own, you'll need to find these parts to build a Visor cable:

- ❑ Three conductor flat cables with crimps (available from www.mouser.com; cable and crimps separate).
- ❑ Molex three conductor housing (available from www.mouser.com).
- ❑ Visor hot sync connector kit (www.atlconnect.com/thirdparty/handspring/pricing/connectors.htm, part #500835).
- ❑ Sonic screwdriver for putting it all together. If you don't have a sonic screwdriver, you'll need a soldering iron and solder.

Soldering the Connector and Printed Circuit Board

The kit contains a black connector, shown here, that provides the connection to the Visor. The connector has eight metal connector pins. On each side of the metal pins are plastic pins that will be used to align the connector to the circuit board.

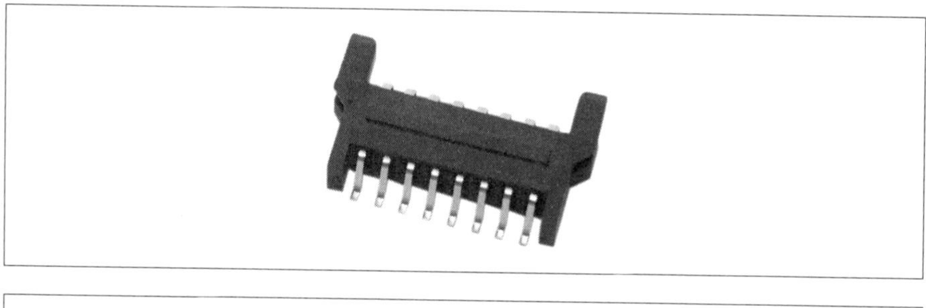

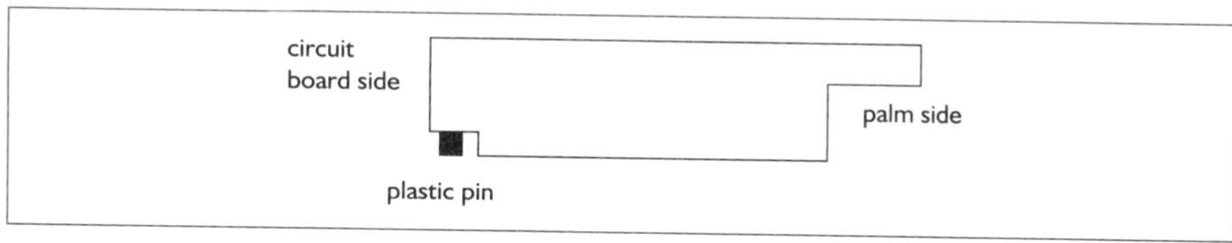

The Visor connector contains a printed circuit board with a small notch cut into each of the short sides, as shown next. Also, each connection on the board shows a printed number just to the left of the connection. These numbers will come in handy later when we solder the connector pins to the circuit board. (They also come in handy when you are trying to remember how many dwarves there are in the story of Snow White. Yes, there was an eighth dwarf; his name was Geeky.)

Position the black connector onto the circuit board. The black pins on the connector should fit into the notches on the circuit board. The connection should be just snug enough to keep the two parts together, as shown here:

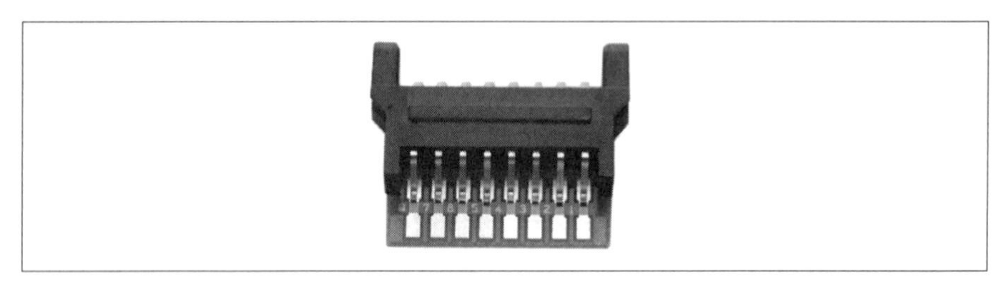

Solder pins 1, 4, and 8 to the circuit board, as shown next. (Thinking about soldering the other connections? Don't. They won't be used, and soldering them simply increases the chances of creating a short circuit among the pins that are used.)

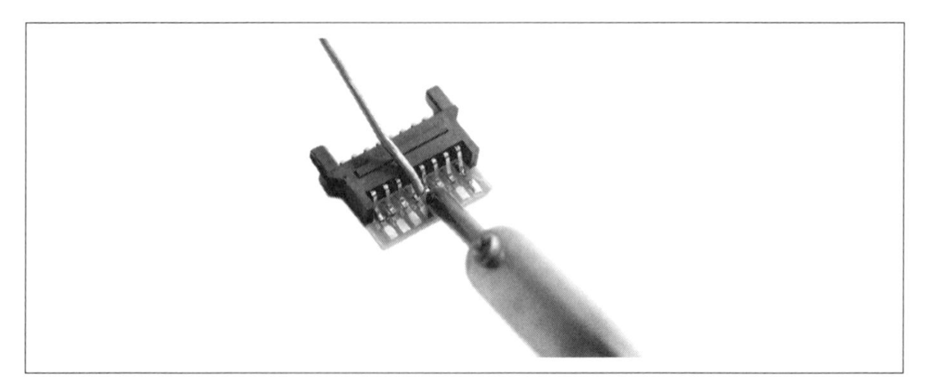

The Safety Geek Says

Both the previous and next steps involve some soldering. If you've built electronic circuits or dabbled in hobbies like radio control models, you're probably already familiar with how to solder—more importantly, how to do it safely. If you've never soldered a connection before, then *STOP!* Find someone who knows how to solder and ask for a quick lesson on soldering connections.

When soldering circuit boards, follow these tips:

- The pins are close together, and pins 1 and 8 are close to the plastic of the connector. Be careful to heat only the connection you are trying to solder, and don't melt the plastic.
- Use as little solder as possible. After soldering the three pins, examine the board carefully to ensure you have not created a short circuit with any of the other connectors.
- Cool your new connection as soon as you complete the solder. We keep a damp rag nearby to dab the solder right away. If you let it sit too long, the extreme heat of the solder can damage nearby circuits or connections.
- Solder contains lead, and lead is toxic to humans like us. Take care to wash your hands after using lead solder, and avoid eating the circuits.
- Don't inhale the fumes from the soldering process.
- The soldering iron, hot enough to melt solder, is also hot enough to give you a nasty burn. Treat the iron with care. Unplug it when you are finished.
- The connection you just soldered is as hot as the iron. Give it a chance to cool before you touch it.

Connecting the Cable

Now it's time to solder the cable to the connector. The cable that comes with the kit has crimps on one end and bare wire on the other end, as shown next. If the other end is not stripped, separate a short section of the strands and use a wire stripper to strip a short (1/8 inch or less) section of each wire.

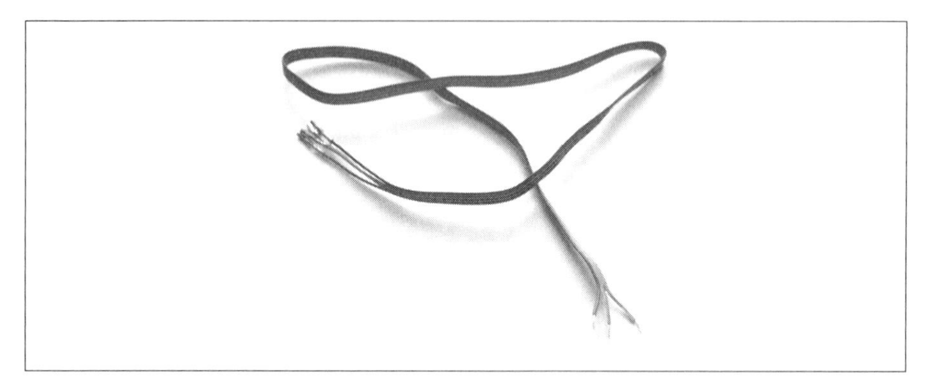

The wire on one side of the cable will have a red stripe painted on it. Solder this wire to circuit board connector 4. Solder the wire on the other side of the cable to the circuit board connector 8. Finally, take the middle wire and solder it to the circuit board connector 1. If you have a four-wire cable, you can do it the hard way like we did and solder the wire next to the red wire to connector 1 and the outside wire to connector 8; then clip the other middle wire and leave it unsoldered. Or you can simplify things by pulling the outside wire off the cable, leaving a three-wire cable.

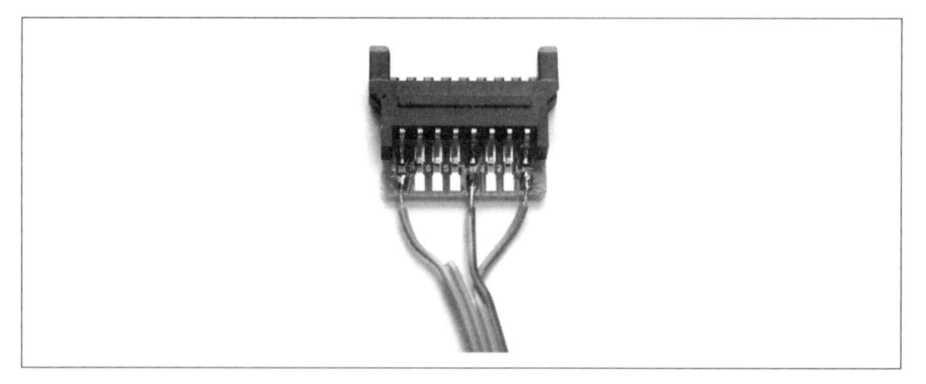

The Safety Geek Says

A short circuit can turn your world-conquering robot into just another pile of electronic junk. Carefully check the solder connections you just made to ensure that no adjacent connectors are soldered together.

The black Molex connector will have four slots. It will also have a small arrowhead imprinted on the connector. Find that arrowhead and insert the red wire into the slot that lines up with the arrowhead. Insert the middle wire into the slot next to the red wire. Skip the third slot and insert the final wire into the slot on the right side of the Molex connector.

If you still have a four-wire connector, insert the wire next to the red wire into the slot next to the red wire and insert the third wire into the third slot. When connected correctly, each wire in the cable will be inserted in order into the connector with no wires crossing over, as shown here:

Assembling the Visor's Connector

The connector housing kit has two metal clips, shown next, and two black plastic buttons. Each button has slots that slide over the metal clip. Slide each button over one of the clips.

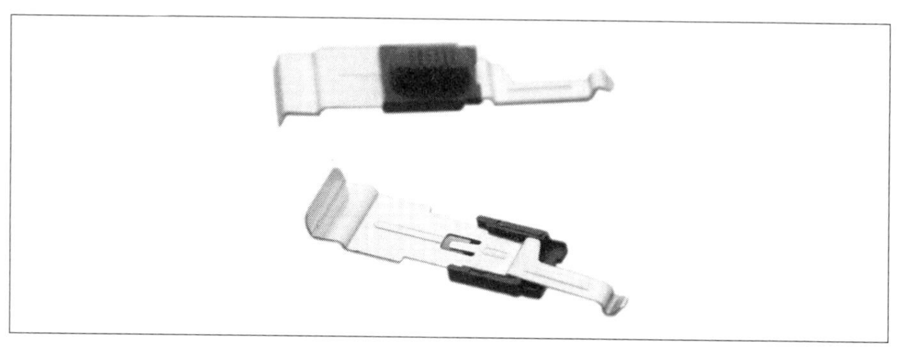

Take one of the connector housing sides. Insert both clips into the connector housing. The metal clips fit into L-shaped slots in the housing. Also, when positioned correctly, the black buttons will extend through a small hole in the side of the housing.

Insert the circuit board and connector into the housing. The metal clips will help hold it in place.

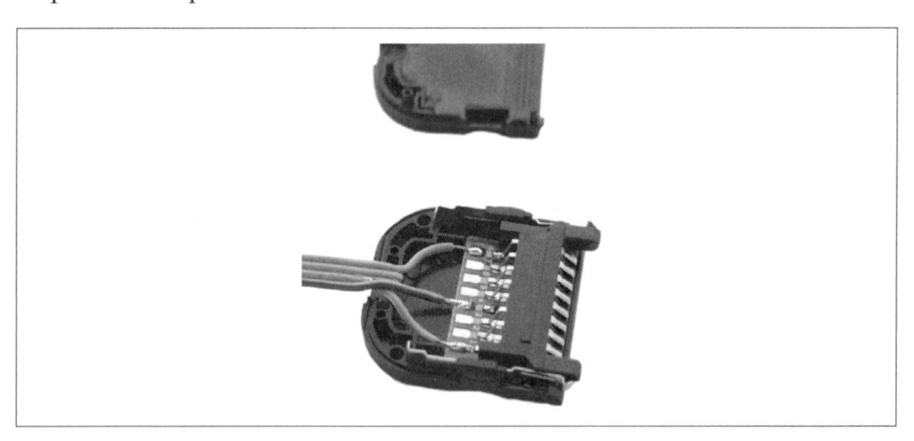

Snap the other housing piece into place.

Your Visor BrainStem connector cable is now complete. You can skip the rest of the chapter and go on to the next chapter.

Using a Handspring Visor with the PPRK

Famous quotes that got it wrong:

- "The world is round, and we'll be in China in just a few months."
- "They're just a few colonists with no army, so we'll have no problem quelling that revolution."
- "...any handheld computer (such as Palm, Visor, etc.) that supports serial communication should be usable for PPRK."

To be fair, that last quote, from the Carnegie Mellon PPRK web page, starts with the phrase, "We tested only Palm III, but..."

The folks at Carnegie Mellon knew that the PPRK might not work with other PDAs. As it turns out, the Handspring Visor doesn't work quite as well as the Palm III or Palm V. To use the Visor with the Pontech controller you'll need to make a connector cable as well as a nontrivial modification to the controller board.

The Robot Geek Says

Top Ten Reasons You Should Buy a BrainStem PPRK

- We really, really, really (really!) recommend that if you have a Handspring Visor, you get the BrainStem kit. Really.
- You can modify your SV203 controller to work with the Visor, but this is definitely a task for a hardcore roboticist.
- You will need to build your own connector cable.
- Did we mention that we recommend the BrainStem controller for the Visor?
- You can use the Visor with a BrainStem without any modification to the BrainStem.
- If you choose to modify the SV203, you do so at your own risk.
- Modifying the SV203 could possibly lead to a reconfiguration of the galactic dimensional matrix, resulting in a universe where jazz is still the most popular genre of music in the world. Dave thinks that situation is akin to the end of the world.

- You can also damage your SV203, making it unusable.
- We cannot be responsible for damage you cause to the controller, to other components, or to the universe, as a result of attempting this modification.
- Really, if you have a Visor, consider using the BrainStem kit.

Parts List

Pressing on? Here's what you'll need:

- Visor hot sync serial cable, or Visor hot sync connector kit (www.atlconnect.com/thirdparty/handspring/pricing/connectors.htm, part #500835, and three-wire cable)
- DB9 male connector (www.mouser.com or Radio Shack, or other electronic supplier)
- Ohmmeter or other circuit checking device (Radio Shack or other electronic supplier)
- Pontech SV203 controller (www.pontech.com, or www.acroname.com)
- Soldering iron or sonic screwdriver
- Solder wick or other solder removal tool (Radio Shack or other electronic supplier)

If you have a Visor serial cable with a DB9 female connector, skip ahead to the section "Connecting the Cable to a DB9 Male Connector." If you are building your own cable, start with the next section, "Building a Cable."

Building a Cable

If you are using the Visor hot sync connector kit to create the cable, follow the directions in the previous section for soldering the connector and the cable and putting the connector together (but not the part about inserting the crimps into the Molex connector). In a nutshell, those steps go like this:

1. Position the connector on the circuit board and solder pins 1, 4, and 8.
2. Solder the three wires in the cable to the circuit board—the red wire to connector 4, the middle wire to connector 1, and the remaining wire to connector 8.
3. Slide the buttons over the metal clips.

4. Insert the metal clips into the housing.
5. Insert the circuit board into the housing.
6. Assemble the connector.

If all goes well, at this point, you will have a cable with a Visor connector at one end and unattached wire at the other end, as shown here.

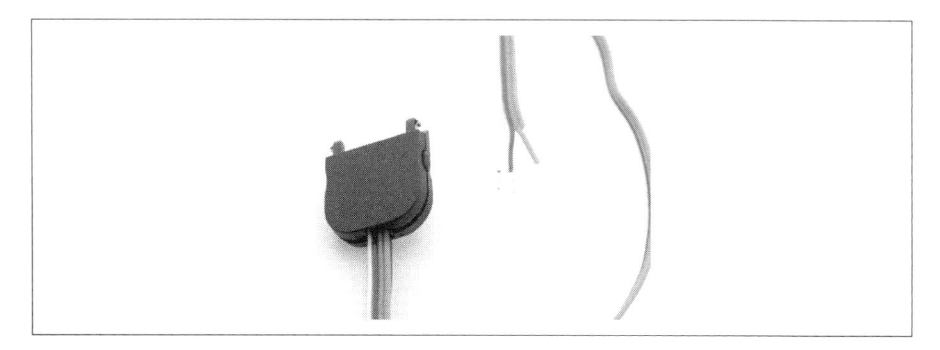

Connecting the Cable to a DB9 Male Connector

Start by cutting the cable approximately 18 inches from the Visor connector. If you have an actual Visor serial cable that you are modifying, you will need to find the wires in the cable that correspond to pins 1, 4, and 8 in the Visor connector. If you built your own connector, you may already know which wire is which, so you can go on to the next step. Otherwise, set your ohmmeter (shown next) to check for resistance, and by touching the ohmmeter probe to pin 1 and to each of the wires at the other end of the cable, determine which wire is connected to pin 1 of the connector. Label wire 1. The pins of the connector are numbered from right to left. Follow the same procedure to identify the wires connected to pins 4 and 8 of the Visor connector.

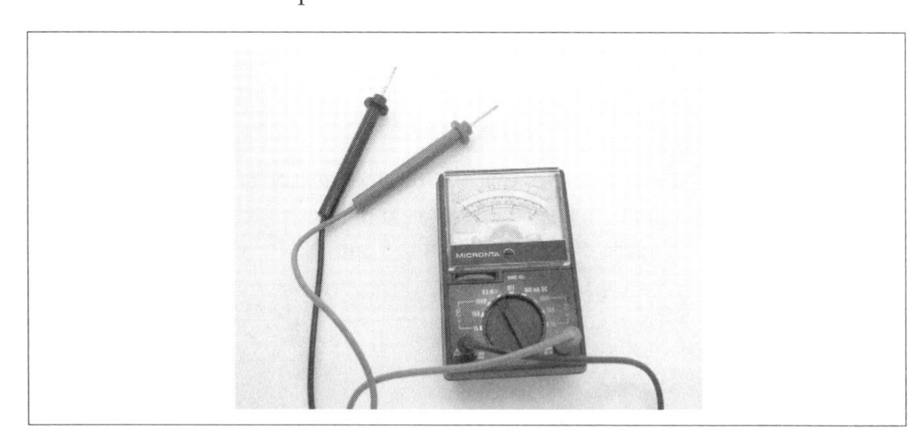

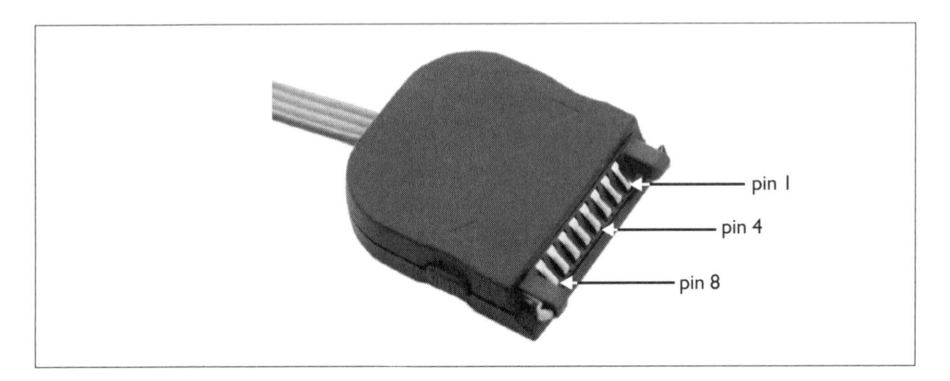

The Robot Geek Says

A Cheap Connection Tester (Based on Evan Johnson's Sixth Grade Science Project)

If you don't have an ohmmeter, you can easily create a simple continuity checker to determine which wire is connected to which connector pin. Try this:

- ✖ Create a simple electric circuit with probes or clips that can be attached to each end of a conductor. The circuit consists of a simple serial connection between batteries and a light bulb with some break in the circuit. Each end of the break is a "probe" that can be used to check for circuit continuity.

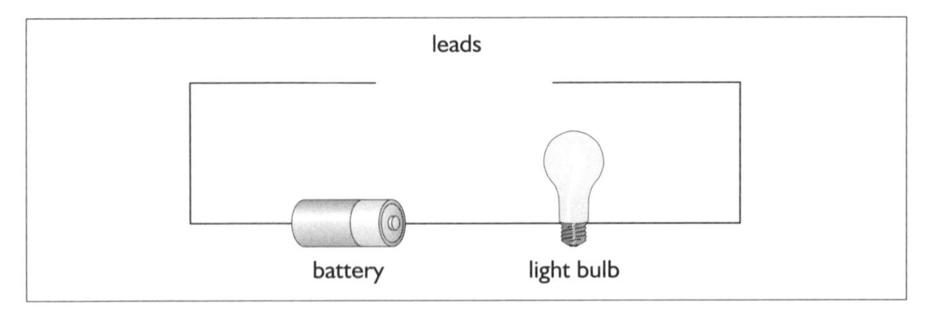

- ✖ By touching one probe to a connector pin, each wire can be tested for conductivity. When the two probes touch the correct pin and wire, the electric circuit is completed and the light bulb will light.

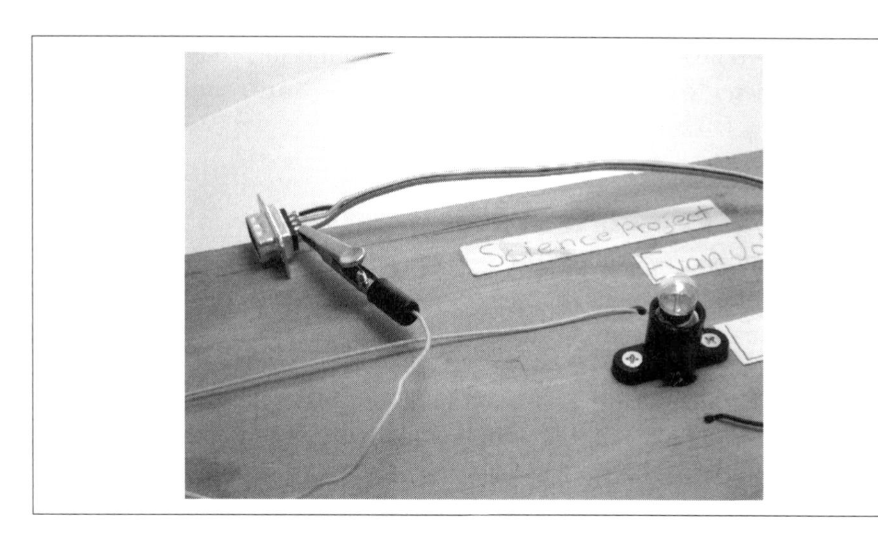

Next, solder (or otherwise attach) each of the three wires to the DB9 connector as shown in the table. Looking at the back of a DB9 male connector, the pins are numbered from right to left on the top row. The rightmost pin is number 1, and the leftmost is number 5.

Visor Pin	DB9 Pin
1	2
8	3
4	5

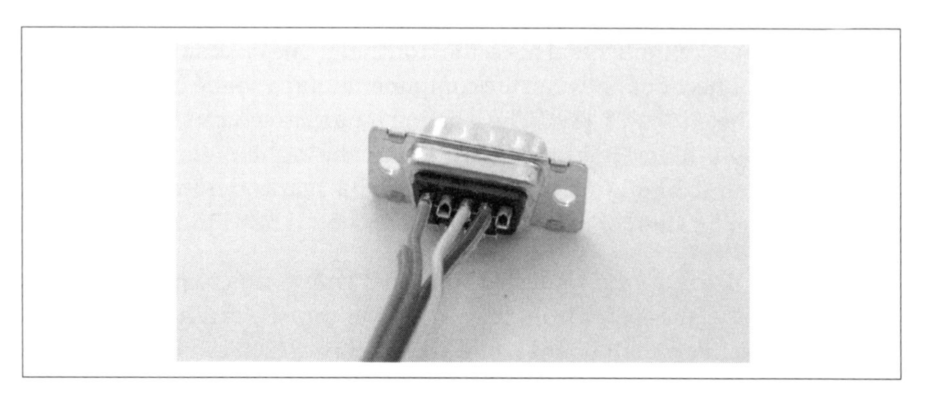

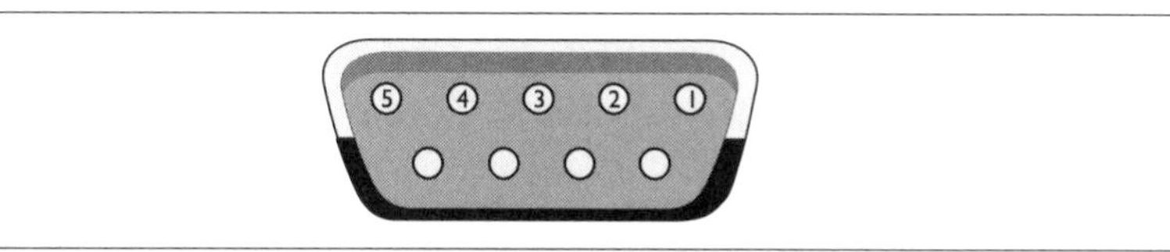

If necessary, reassemble the DB9 connector. You may want to use the ohmmeter to double check that pins 1, 4, and 8 are connected to pins 2, 3, and 5 of the DB9 connector.

Modifying the Pontech SV203

Now it's time to modify the Pontech SV203 controller to work with the Visor. This requires that you remove one of the chips from the controller and make some new connections on the board. This is *not* a trivial task. We don't mean to scare you away from trying it, but just be advised that it's a project that is best left for the more advanced hobbyist. Dave, for instance, shouldn't try this. Kevin, on the other hand, is actually salivating at the thought of prying this puppy open.

The Robot Geek Says

Reasons Not to Modify the Pontech SV203

- Modifying the Pontech SV203 controller board will almost certainly void your warranty, could damage the controller, and might, as we've mentioned, lead to a reconfiguration of the galactic dimensional matrix.
- Did we mention that we recommend you use the BrainStem controller with a Visor?

If you examine the SV203, you'll find one side with connector pins and various components. This is the front side. The back side consists of the electrical connections between the components and a single IC (integrated circuit) chip labeled "MAX 489." The labeling is hard to see, but trust us, it's there; you can see it if you hold the board at just the right angle. Also, you'll see a small, round dimple in the chip. That dimple marks the location of pin 1. Remember the location of pin 1, because later you'll need to know where it is.

Heat up the soldering iron: that chip needs to go. Carefully melt and remove the solder from the pins of the chip. Be careful not to overheat or damage the circuit board while you do this, which can happen if you continue to apply more and more heat to the board. We recommend that you melt one pin at a time and immediately cool the board after each pin is melted. Then reapply the soldering iron to work on the next pin, cooling it immediately as soon as the pin is gone. Work your way around the chip this way.

The Robot Geek Says

Methods for Removing Solder

You can use a number of tools for removing solder from a connection. We recommend these:

- **Suction** This involves using a device that sucks up melted solder. Use a tool designed for that, not a straw and your lips (Dave learned this the hard way). Radio Shack #64-2098 "Vacuum Desoldering Tool" is one choice.
- **Solder wick** This consists of a braid of wire that wicks up molten solder from a connection.

Gently remove the chip from the board. You may need to use a small Exacto knife or a similar sharp instrument to remove the chip. Again, take care not to damage the board. After the chip is removed, examine the circuit board to ensure that no stray solder has short circuited any of the connections. If you see any stray solder, carefully clean it off the board with your Exacto knife or with the soldering iron and solder removal device.

Using fine-gauge wires, create jumper connections between pins 2 and 11, and between pins 5 and 10. To find these pins, count the pins moving in a counterclockwise direction from pin 1. Moving right to left from pin 1 are pins 2 through 7. Immediately below pin 7 is pin 8; the bottom row of pins are numbered from left (pin 8) to right (pin 14). Again, check for short circuits when you are finished, removing any that you find.

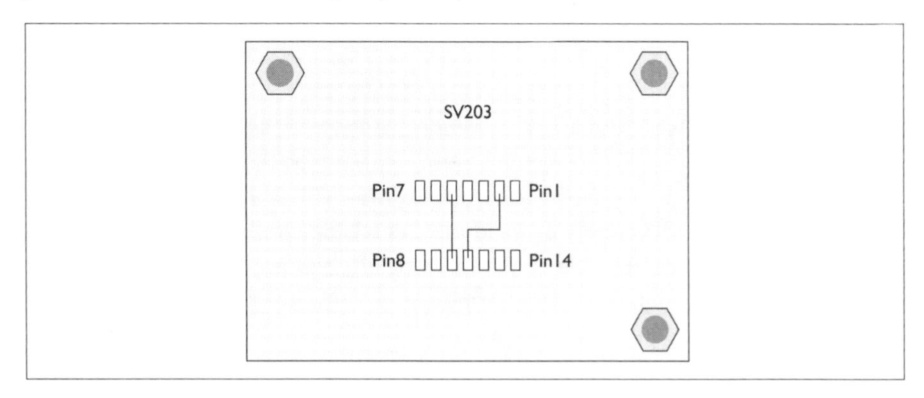

The Pontech controller is now ready to use with your Visor. You can skip the remainder of the chapter and go on to the next chapter.

Using a Palm VII with the PPRK

As we mentioned at the beginning of this chapter, the Palm VII has the same style of hot sync connector as the Palm III, so it will work with either the Pontech or BrainStem controllers—at least, it should work just fine in most cases. Consequently, before proceeding with this section, you should skip ahead to Chapter 5, download some of the sample programs, and attempt to use them with your Palm VII. If, after attempting to run a program, your Palm VII does not seem to be communicating with the SV203, you should come back to this section and use the troubleshooting information here.

If your robot does not seem to be working, first check that all the connections are correct. Check the wiring between the servos and the controller and the sensors and the controller. Ensure that each Molex connector is plugged in the board in the correct orientation. If the robot appears to be constructed correctly, you may need to check the communication link between the Palm VII and the SV203.

Some Palm VII devices will have problems communicating with the Pontech controller because the speed at which the Palm device is talking does not match the speed with which the SV203 is listening. You can fix that problem by adjusting the communication speed of the Pontech controller.

The Robot Geek Says

Communicating at Nearly the Speed of Light

When two electronic devices are communicating, they need to speak the same "language." The Pontech accepts commands over an RS232 serial connection. When the Palm VII sends a message to the SV203, or the SV203 sends a sensor reading to the Palm VII, the message needs to be formatted according to the RS232 format. Part of that format includes the speed, or baud rate, at which data is sent over the wire. If the two devices are communicating at different speeds, they won't be able to understand each other.

The baud rate at which the SV203 operates is set by a memory location in the EEPROM of the SV203. The controller can be set to any baud rate between 2400 to 19200 baud. The default setting is 9600 baud, which is the same rate as the Palm VII. However, if either device is slightly off 9600 precisely, the Palm VII will not be able to communicate with the SV203. Symptoms of this problem include these: the Palm sends commands that the SV203 does not seem to receive, or the Palm VII does not receive sensor readings from the SV203.

To solve the communication problem, you will need to change the baud rate of the SV203. And that leads us to the joke of the day:

Q: How many roboticists does it take to change the baud rate?

A: None; it's a software problem!

This, of course, means you'll need to take a look at Chapter 7 to learn how to write a program for the Palm VII and the PPRK, and you'll use that program to adjust the baud rate of the SV203. If you're not that ambitious, we provide a program for you in the next section. (This is starting to feel like one of those interactive, reader-directed stories. If you choose to open the chest, go to page 395. If you choose to go through the door marked "Danger! Here be dragons." go to page 277.)

If you need to change the baud rate, you will need to start at Chapter 7, where you can see how to write and use programs for the Pontech controller. Read the section about using BASIC programs. Go ahead. We'll wait here patiently.

Back already? Okay, let's go.

Changing the Baud Rate

If you check the SV203 technical manual that came with the controller board, you'll see a section on commands you can send to the SV203 to change its configuration, to read a sensor, or to send a command to a servo. The command we are interested in is the command to change the EEPROM setting for the baud rate. Here's the command to write to the EEPROM memory:

```
WEm n
```

You can find this command detailed on page 20 of the SV203 technical manual. The `m` is an integer that can range from 0 to 11. The EEPROM register for the baud rate is register 9. The n is an integer that can range from 24 (19200 baud) to 200 (2400 baud). The default setting is 50, which sets the baud rate to 9600.

So, the command to set the baud to 9600 is

```
WE9 50
```

To change the SV203 register, you send this command as a string to the controller board. Now, all we need is a simple program to do that. The following program is written for the HotPaw Basic interpreter. The HotPaw interpreter is on the CD-ROM for this book in the \Other Applications directory. If you haven't yet done so, install HotPaw Basic to your Palm VII. (You can find more information on how to do this in Chapter 7.) Here is the program to change the baud rate of the SV203:

```
#chbaud.bas
15 open "com1:", 9600 as #5
20 draw -1
25 draw "Set SV203 baud rate
30 form btn 75,95,70,20, "Send Command ",1
35 form fld 80,55,40,12, "50",1
40 draw "command value (24 to 200) = ",5,42
45 x=asc(input$(1))
50 sd$= "WE9 " + s$(0)
55 print #5,sd$;
60 sound 1800,100,64
65 sound 1500,30,64
70 close #5 :
75 goto 15 : end
```

In line 15, the program opens a communication channel through COM1, at 9600 baud. In lines 25–40, the program draws the form controls.

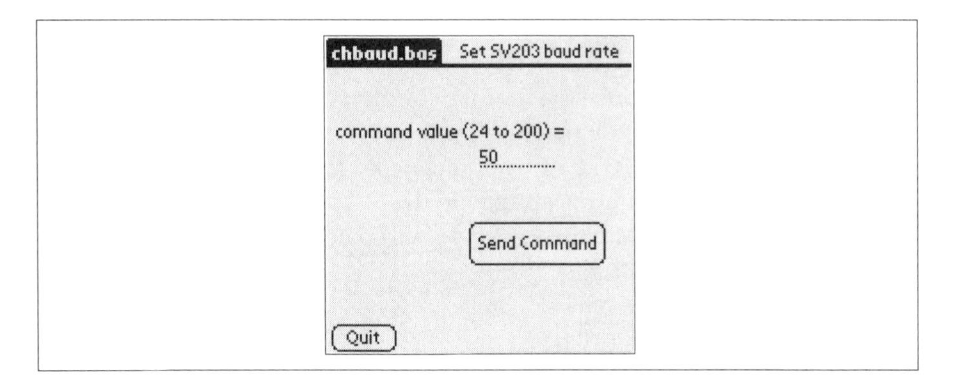

The program then waits for you to enter a value in the form field. The default value is already set for you (50). Enter a new value to be set in the EEPROM and click the Send Command button. The program constructs the command in line 50, and then it sends the command to the SV203. It finishes by playing a little beep and closing the communication channel.

HotPaw programs are saved as memos for the interpreter to run. The easiest way to enter the program is to find it on the CD-ROM with this book, save it as a memo in the Palm Desktop, and then hot sync your PDA. Hot syncing will transfer the memo to the PDA.

Enter and save this program to your Palm VII and attach the Palm VII to the SV203. If needed, connect the power supply to the SV203 and turn the switch to the on position. You can now use the program to set the baud rate of the SV203.

Start by entering values close to 50, such as 49 or 51. After changing the baud rate, try one of the sample programs from Chapter 5 again. If the robot can send and receive from the Palm VII okay, you are done. Otherwise, you will need to try a new value. If you tried 49 the first time, try 51; if you tried 51, use 49 the second time. After each new value, test one of the sample programs. If you need to try additional values, move by increments of 1 away from the default value of 50. Eventually, you should find a baud rate that allows the Palm VII to communicate with the SV203.

Restoring the Default Value

There is a chance that before you find a baud rate that works, you could set the SV203 to a baud rate that prevents the Palm VII from communicating with the SV203 at all. That means, of course, that the BASIC program above will stop communicating with the SV203. At this point, you will need to reset the SV203 manually to the default baud rate setting.

To reset the Pontech SV203 to its default setting, you will need to create a circuit between two pins on the SV203 controller. Start by turning the power switch to the off position. The SV203 has a block of pins marked J3. These are the same pins into which the yellow wires from the sensors are plugged. If pins 7 and 8 in block J3 are connected while the board is powered on, the baud rate of the board will be set to its default value. Connect each end of a single wire to pins 7 and 8 as shown in the following illustration. Switch the power on, and leave the SV203 on for at least 3 seconds.

This will reset the baud rate to 9600 and allow the Palm to communicate with the SV203 once again.

Finishing Up

Once you've found a baud rate that works, write down the EEPROM value you used and keep it in a safe place. You may need it in the future. You are now ready to start playing with your robot.

Chapter 5

Checking Out the Robot

Congratulations! By this point, you should have a completely assembled robot. If you're like Dave, you'll also have a small collection of spare parts that actually outnumber the original components you started with, an incinerated tablecloth from the soldering, and a cat that now needs a considerable amount of therapy. That may not sound good, but you should have seen him try to build his own PC.

So far, so good. Insert the batteries into the battery holder, flip the switch to the on position, and watch that robot go! All hail the mighty robot! Robot, go! Go, robot, go!

Oops, sorry ... not quite yet. You're still missing one vital piece of the puzzle: the robot's software. The software tells your hardware what to do, how to do it, and how frequently it should do it. Without a little programming, your bot simply isn't going to do much. That's why, in this chapter, we'll show you some places where you can get some software that will allow you to start playing with your PPRK (Palm Pilot Robot Kit).

Software for the PPRK

Because of the great interest in the PPRK, a number of web sites have posted sample software for the robot. The two obvious places are the original PPRK web site and Acroname, but a few other sites are out there as well. In this section, we'll look at some places where you can get sample code for the PPRK. Then we'll use some of this sample code to test the PPRK. Although you can download all the code from the web, don't bother; you'll find all the software on the CD-ROM that's attached to the back of this book.

TIP: If you built a robot using the BrainStem controller, you'll want to jump to the next major section of this chapter, "Software for the BrainStem."

Software from Carnegie Mellon

The Carnegie Mellon PPRK web site (www-2.cs.cmu.edu/~pprk) contains a small but useful section of software to get your PPRK up and running. Click the Software link in the left column to get to the software page.

The main software page contains three programs you can download and use with your PPRK. They are ServoTest.prc, PenFollow.prc, and Robot1.prc.

- **ServoTest.prc** This program does exactly what the title implies: it tests the servos. This is a good program to install and run when you're verifying the operation of your new robot.
- **PenFollow.prc** This program moves the robot in response to your moving the stylus across the PDA screen. In other words, the robot follows the direction of pen movement on the Palm screen.
- **Robot1.prc** This program has several routines to control the robot, including following a wall or a target, moving and spinning simultaneously, and displaying the readings from the infrared (IR)

sensors. Unfortunately, Robot1.prc has a bug, and it doesn't report the correct ranger reading for the third ranger. If you really want to verify all three rangers and the servos, check out the Config program, which we discuss later in the section "Software for the BrainStem." Despite its location in the BrainStem section, Config works with the SV203 also.

Since these programs are already compiled into Palm OS applications, using them is as easy as downloading them from the web and performing a hot sync operation to load them to your PDA. Take a moment to do that now with Robot1 and one of the other two programs—we'll wait for you.

NOTE: ServoTest and PenFollow conflict with each other. If you try to install both of them on the Palm at the same time, only one will show up in the Application Launcher.

ServoTest

ServoTest appears on your PDA as a Robot Test icon. Tap the icon and you will see a simple interface that consists of a number of sliders that can be used to control up to six servos. With the PDA attached to the robot, move sliders 1, 2, or 3 (shown next), and the corresponding servo on the robot will rotate.

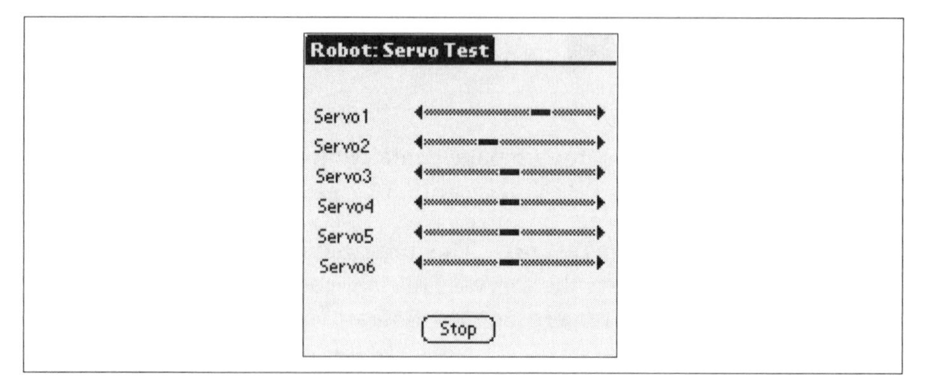

PenFollow

The PenFollow application (which appears rather conveniently as a Pen Follow icon on the PDA) also has a simple interface. It consists of a blank screen with two buttons on the bottom. Tap the Start button, and move your stylus across the screen. A line is drawn from the center of the screen to the current stylus location, and the robot will start moving in the direction of this line. Move the stylus, and the robot will follow it.

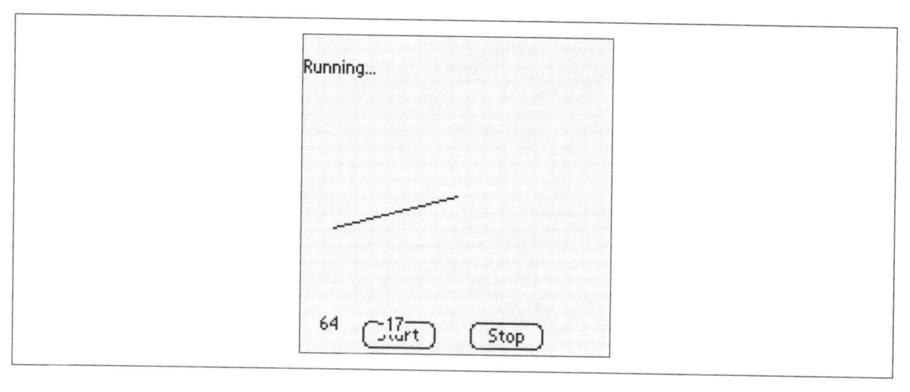

Robot I

This program appears as a Palm Robot icon on your PDA. Tap the icon to run the program and a number of buttons will appear on the screen. To test the target following capabilities of your robot, for example, tap the Target Follow button. Tap that button and see if you can get the PPRK to follow you around the room.

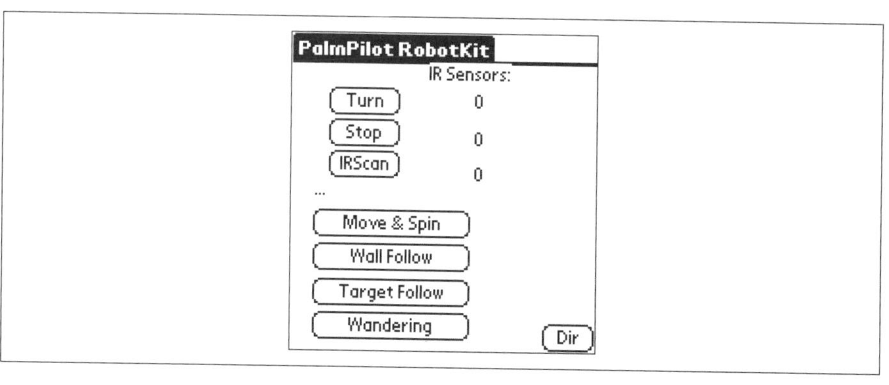

Tapping the other buttons will cause the robot to execute a number of other behaviors:

- Tap Move & Spin to see a demonstration of the robot's ability to move in a line and spin simultaneously.
- Choose Wall Follow and, not surprisingly, the robot will attempt to follow a nearby wall (located using the infrared sensors).
- Tap the Wandering button to put the robot into a mode in which it wanders around the room, avoiding any obstacles it finds.

Other Software from Carnegie Mellon

The designers of the PPRK used the C++ language to write programs for the PPRK, but C++ is not the only coding option. The Palm device controls the robot by sending simple text commands to the SV203 controller. For example, if you look on page 10 of your SV203 technical manual, the command to select a servo is

```
SVn
```

where `n` is the servo number. The command to move the servo (found on page 11 of the manual) is

```
Mn
```

where `n` is a number from 0 to 255.

Any software that can connect and send strings of data over a serial line can be used to control the robot. A number of ingenious people have used other languages to write sample programs for the PPRK. The Carnegie Mellon PPRK software page has a link that will take you to some of these sample programs. Look for the sentence that reads Programs Written By The PPRK Community! Click Here. Click the link to go to the Community Programs page. Alternatively, go directly to this web page: www-2.cs.cmu.edu/~pprk/programs.html.

When we last visited it, the page included a few BASIC programs and a C program to fiddle with; more may be offered by the time you read this.

On this screen, three programs are written in the BASIC programming language for HotPaw Basic, and one C language program is written for PocketC. The three BASIC programs are sequence_motors_test.bas, triangle3.bas, and square2.bas. As you might guess, these programs test the servos, move the robot in a triangle pattern, and move the robot in a square pattern. The C Program is PocketPPRK and it demonstrates a number of different features of the robot.

Using a BASIC Program with the Robot

Let's use one of the BASIC programs to test out the PPRK. Before you can do that, though, you'll need to get a BASIC interpreter for your PDA. After you've loaded an interpreter to the PDA, you can load and run BASIC programs on your PDA. So let's start with the interpreter.

Loading HotPaw Basic

Several BASIC interpreters are available for the Palm OS. The three programs on the Community Programs page were written for HotPaw Basic. We've included a copy of this application on the CD-ROM to make it easy for you to get started. Find the ybas138b0.zip file in the \Other Applications\HotPaw directory of the CD-ROM and unpack it to your PC. You need to load two files to your PDA: ybasic.prc and mathlib.prc.

The Robot Geek Says

Basic and Palm OS Devices

- Even though we used HotPaw Basic to run the sample program here, you can write programs for your robot using any BASIC interpreter that runs on a Palm OS device. In addition to HotPaw Basic, we've included NS Basic on the CD-ROM.
- For now, you don't need to worry about how to write a BASIC program or how it executes. If you're really interested, we cover that topic in Chapter 7.

HotPaw Basic looks for program files in the Palm's Memo Pad. Program filenames begin with a pound (#) sign, are followed by the name of the program, and end with a *.bas* extension. Here's how to prepare your PDA to run the program:

- ❑ Download the triangle3.bas program (available at www-2.cs.cmu.edu/~pprk/programs.html or on the CD-ROM.)
- ❑ Open triangle3.bas in a text editor and copy it into a memo in the Palm Desktop on your PC. If you synchronize with Microsoft

Outlook—not Palm Desktop—copy the program into an Outlook Note instead.

- ❑ Perform a hot sync operation; the memo will be transferred to the PDA.

With the program now stored as a memo on your PDA, you can run the program using HotPaw Basic. Find and launch HotPaw Basic from the Application Launcher (it will be listed as *ybasic*). This brings up a screen that lists the programs that HotPaw Basic found in the memo pad.

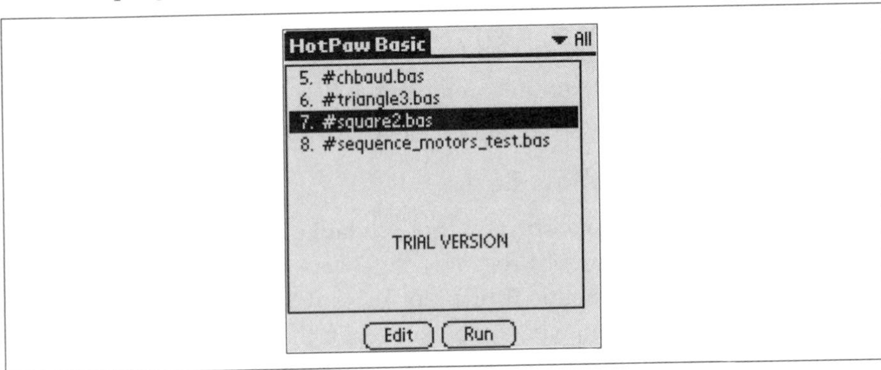

If triangle3.bas does not appear in the list, check to make sure that

- ❑ The program did indeed get loaded to the PDA.
- ❑ The first line of the program begins with a # symbol.
- ❑ The first line of the program then contains the name of the program, which ends with the *.bas* extension.

Tap the program name, #triangle3.bas, and then tap the Run button. The robot should now move in a triangular pattern.

Using a PocketC Program with the Robot

The C program, PocketPPRK, is probably the most extensive program available on the Carnegie Mellon web site. But to use it, you will need to download and install the PocketC runtime for your PDA. Here's what you need to do to run the program:

1. Find the ZIP file PocketC_rt.zip on the CD-ROM in the PocketC directory (the program is also available in a number of places on the web, including www.orbworks.com and www.PalmGear.com).
2. Unzip the archive and install PocketC.prc and mathlib.prc to your PDA.

3. Find the PocketPPRK.zip file on the CD-ROM (or at www-2.cs.cmu.edu/~pprk/programs.html) Unzip the archive and install pprk.pdb and PToolboxLib.prc on your PDA.
4. Find the PocketC icon in the Application Launcher and run it.
5. Select pprk from the list of compiled programs and tap the Execute button.

You will see buttons for seven robot programs. Tapping a button will cause the robot to execute the related program. Here's what the programs do:

- **Wanderer** The robot wanders around, avoiding obstacles.
- **Wall Follow** The robot follows a nearby wall.
- **Spin Follow** The robot follows a wall—while spinning.
- **Leap Frog** The robot follows a wall in a leap frog fashion.
- **Holonomic** The robot demonstrates its ability to move in any direction without turning, using three wheels that are not aligned.
- **Memo** The robot executes a pattern stored as text commands in the memo pad. A number of sample patterns are included in the PocketPPRK.zip file.
- **Scribble** The robot follows a pattern that is doodled on the PDA's screen.

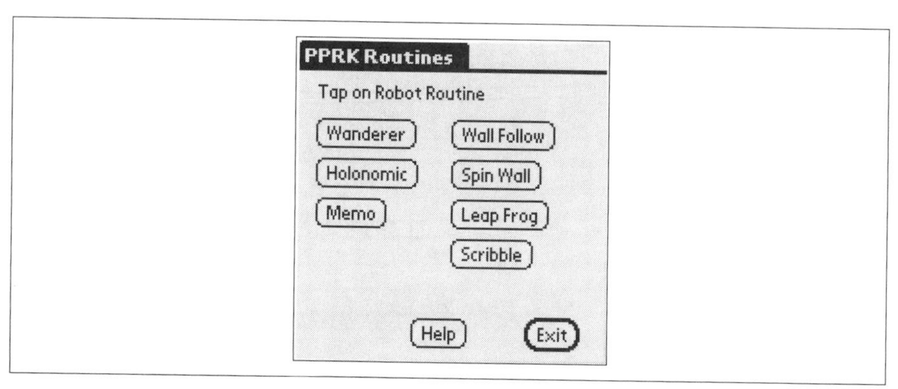

Software for the BrainStem

Your BrainStem controller will not use the same software files as the Pontech SV203. BrainStem is an Acroname product, so most of the software that is available will be from Acroname. Also, because the BrainStem supports many different platforms, you can find software for Windows, Mac, and Palm OS devices, just to name a few.

To make your life easier, all the files from Acroname are already included on the CD-ROM with this book. Simply locate the file you need and unpack it to your hard drive. In the event you've lost the CD, or you want to check for the latest versions of the files, we'll show you how to do that next.

Acroname is continuously updating its software. During the time this book was written, the software was updated from version 14 to version 15. If you want to get the latest BrainStem software, start by going to Acroname's web site at www.acroname.com. Click the Downloads link in the About Robotics set of links, as shown here:

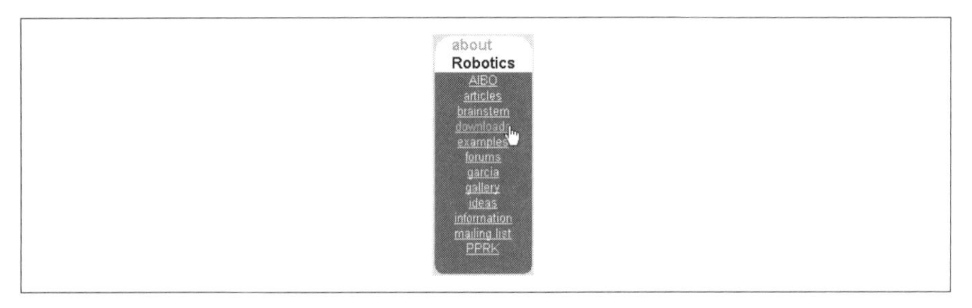

Before you can download any software, Acroname requires that you create a login account. After entering the required information, as shown next, you will be able to navigate to the actual downloads page. (To the best of our knowledge, Acroname doesn't do anything nefarious with your login information. If you are concerned, you can read the privacy statement on the web site.)

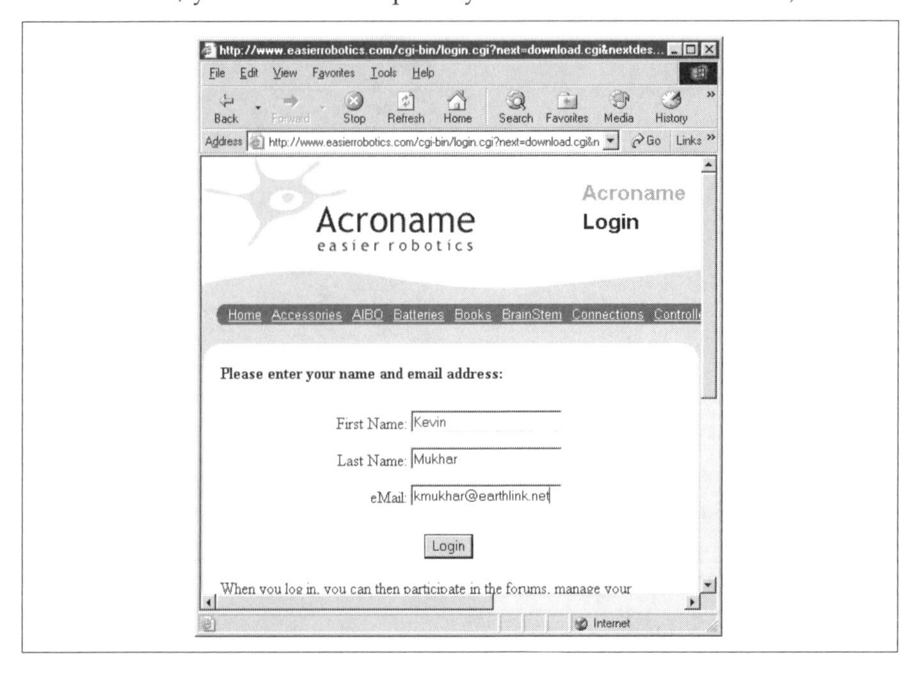

For this chapter, the only file you need is the PPRK SDK file for your platform. For example, if you are using a Palm OS device and a Windows PC, you will download the BrainStem PPRK SDK for the target platform Palm OS and the download platform Win32. *This is important!* Getting the two platforms correct is vital. For example, if you download the Palm OS PPRK SDK for the Mac OS X and you have a Windows machine, the software will not work. Some computers have even been known to go on strike if you try to give them the wrong software.

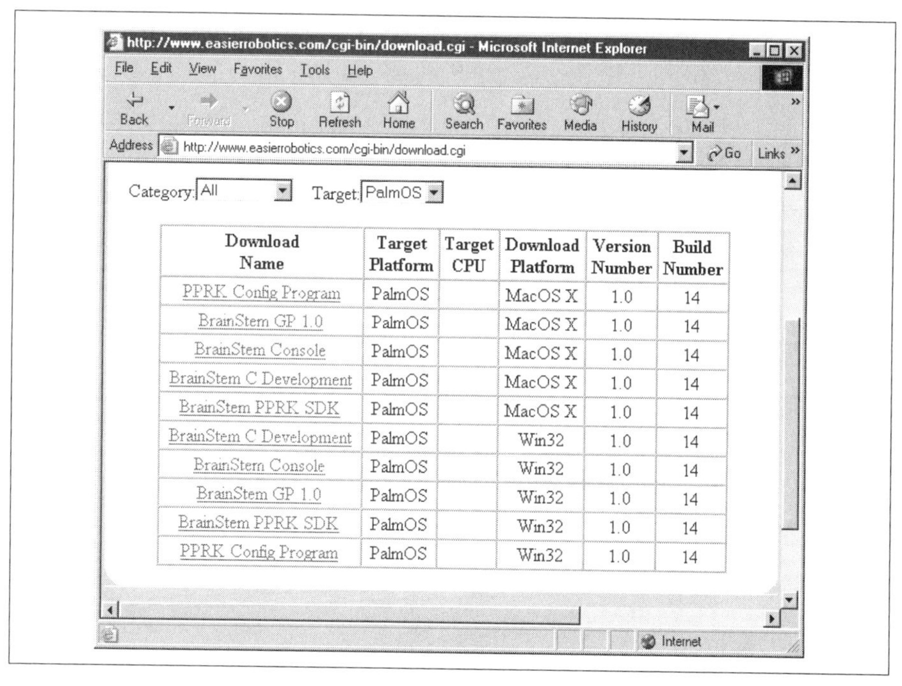

Download Name	Target Platform	Target CPU	Download Platform	Version Number	Build Number
PPRK Config Program	PalmOS		MacOS X	1.0	14
BrainStem GP 1.0	PalmOS		MacOS X	1.0	14
BrainStem Console	PalmOS		MacOS X	1.0	14
BrainStem C Development	PalmOS		MacOS X	1.0	14
BrainStem PPRK SDK	PalmOS		MacOS X	1.0	14
BrainStem C Development	PalmOS		Win32	1.0	14
BrainStem Console	PalmOS		Win32	1.0	14
BrainStem GP 1.0	PalmOS		Win32	1.0	14
BrainStem PPRK SDK	PalmOS		Win32	1.0	14
PPRK Config Program	PalmOS		Win32	1.0	14

The PPRK SDK contains most of the files you need to work with the robot. However, we will be using some of the other files later in the book, so you can download and install those programs now if you wish. The other files you may want are listed here:

- **BrainStem Java Access** This is used for programming the robot using the Java programming language.
- **BrainStem C Development** This is used for programming the robot using the C programming language.
- **BrainStem Reference** Here is documentation for using and programming the BrainStem controller.
- **BrainStem SDK** This contains everything except the BrainStem Java Access files and the BrainStem reference files.

Working with the Robot

After unpacking the software, you can start trying out some of the applications. Acroname supplies several applications for working with the BrainStem controller and the robot. The Config program provides a way to test the analog sensors and servos of the PPRK. The GP program is similar to the Config program, but it's a more general purpose program. Finally, the Console program provides a way to load and launch programs with the BrainStem. All three of these programs are located in the \aBinary directory of the Win and Mac distributions. If you have the Palm OS versions, the files you need will be in the \brainstem directory.

If you simply want to test that your robot is operating correctly, you should run the Config or GP application. To execute a program for the robot, run the Console application. Note that because all three applications communicate with the BrainStem using the same connection, they cannot be run at the same time; only one of the three applications can be running at any given time.

Config

The PPRK Config program is closely tied to the PPRK. The Config application allows you easily to check the function and orientation of the servos and rangers on the SV203 and Brainstem PPRK robots.

To install the Config application to your Palm OS device, all of these five files need to be loaded to the PDA:

- Config.prc
- aIO.pdb
- aPPRK.pdb
- aStem.pdb
- aUI.pdb

Once all five files have been transferred, start the Config application by tapping the Config icon on the Palm's Application Launcher screen. For the Windows or Mac version of the application, launch the Config executable from the \aBinary directory.

NOTE: It is entirely possible that later versions of the Acroname software will have different or more .pdb files. If you download Config from Acroname, and it does come with different .pdb files, load them all to your PDA.

You can see that the PPRK Config displays a graphical representation of the PPRK in Figure 5-1 for the Palm OS version of the program and Figure 5-2 for the PC version. Radio buttons appear for each of the servos, and a colored triangle represents the field of view of the sensors.

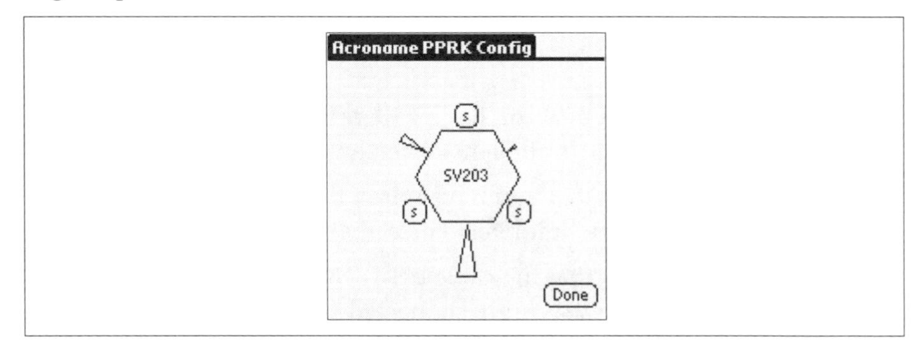

Figure 5-1 The Config program running on a PDA

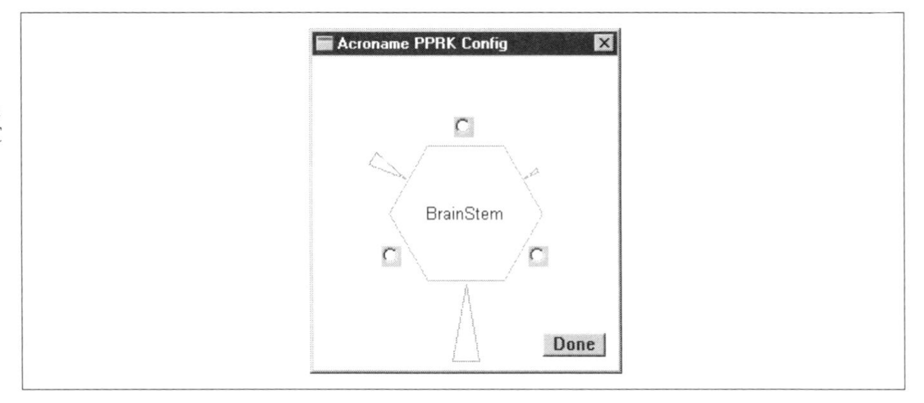

Figure 5-2 The Config program running on a Windows PC

If you select one of the radio buttons for the servos, the corresponding servo on the robot should start rotating.

If you place an object in front of one of the sensors, you should see the corresponding triangle in the graphical display change shape to illustrate that it is sensing an object.

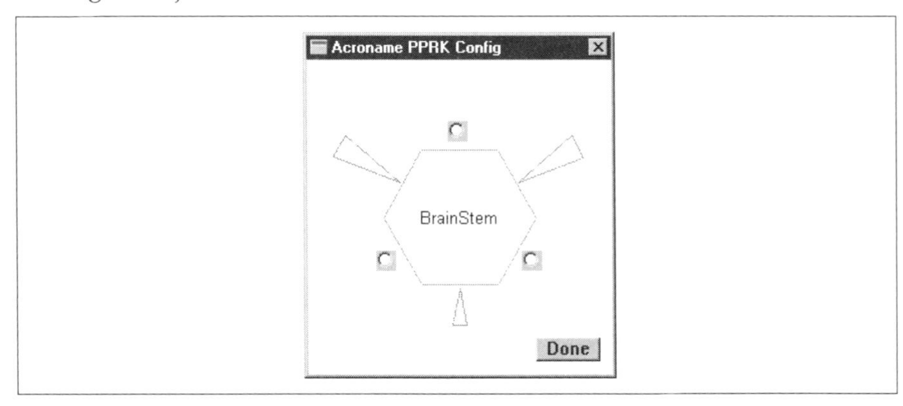

Obviously, if the servos don't rotate when you select a radio button, or if the graphical display doesn't change when an object is in front of a sensor, you have a problem with your robot. Here are some troubleshooting tips for your robot:

- Check that you have wired the connections between the sensors and the BrainStem correctly.
- Check that you have wired the connections between the servos and the BrainStem correctly.
- Check that the cable between the BrainStem and the host PDA or PC is securely connected.
- Check that the power connections are correct.
- Ensure the switch is in the on position.
- Ensure the batteries are good.

GP

GP is a general purpose program for testing the BrainStem controller. Using the GP application, you can see the output from each of the five analog ports, test the input and output from each of the five digital ports, and test and configure any of four servos that could be attached to the BrainStem. Since the PPRK has only three servos, you will use only three of the four in the GP application.

To install the GP application to your Palm OS device, locate the following files. All four of these files need to be loaded to the PDA:

- GP.prc
- aIO.pdb
- aStem.pdb
- aUI.pdb

NOTE: These .pdb files are the same files used by other BrainStem applications. If you've already installed the .pdb files for the Config application, you do not need to install them again.

Once all four files have been loaded, start the GP application by tapping the GP icon on the Palm's Application Launcher screen (see Figure 5-3). For the Windows or Mac version of the application, launch the GP executable from the \aBinary directory (see Figure 5-4). While the GP application is running, neither of the other applications, Config or Console, can be running.

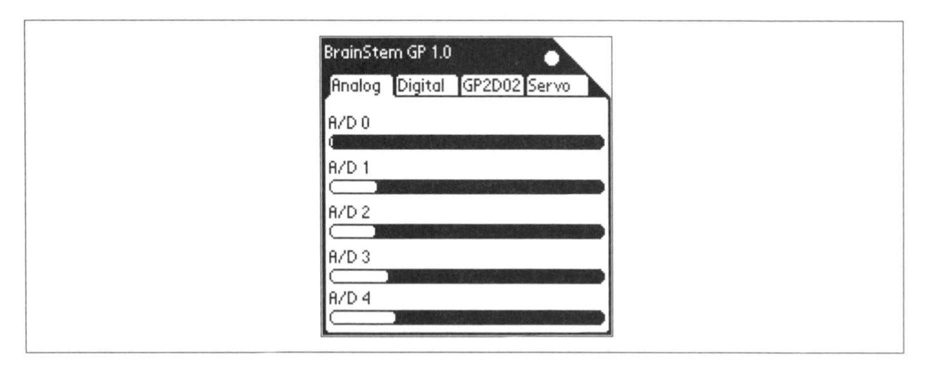

Figure 5-3 The GP application running on a PDA

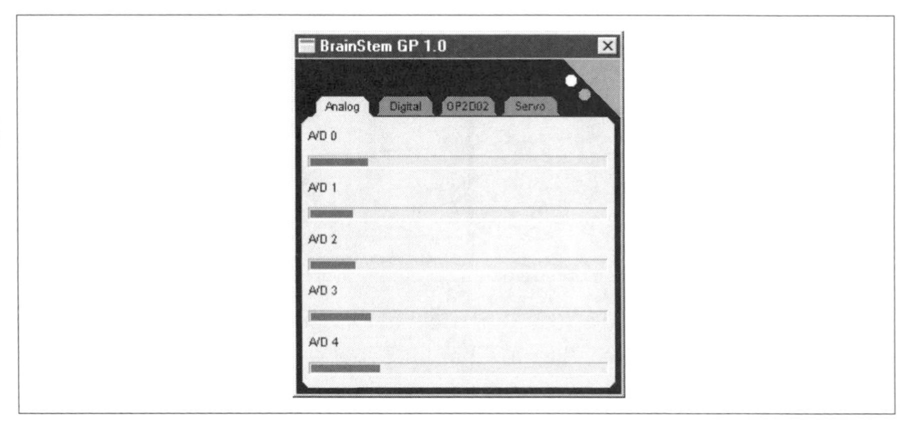

Figure 5-4 The GP application running on a Windows PC

The application needs to make a connection to the BrainStem controller to work properly. For the two devices to connect properly, you must start the GP application first. After it is running, turn on the PPRK. The two LEDs on the BrainStem controller will light up, and within a few seconds, the green light on the BrainStem will start a rhythmic beat, known as the *heartbeat*. The GP application has a similar icon that will flash in sync with the heartbeat. This indicates that the application and the BrainStem are communicating.

Checking the Analog Ports

When you start the application, you will see four tabbed panes labeled Analog, Digital, GP2D02, and Servo. The application starts at the first tab. If you previously started the GP application and changed tabs, select the Analog tab now. You will see five horizontal bars labeled A/D 0 to A/D 4. With the PPRK turned on and connected to your Palm or PC, place your hand in front of one of the sensors. You should see the corresponding horizontal bar change size, depending on how far your hand is from the sensor. Now check the other two sensors. If everything is working properly, the application should show a response from each of the sensors.

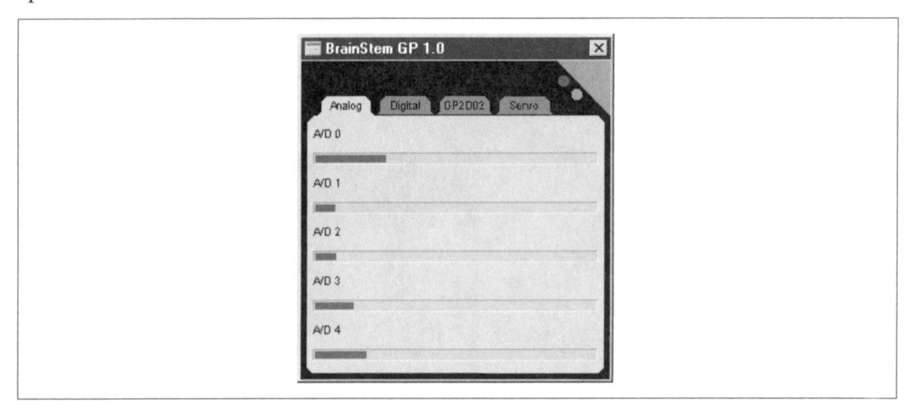

You should also make sure that the sensors are connected in the correct configuration. Recall from Chapter 3 that sensor 1 is opposite the BrainStem controller and closest to the battery pack. Looking at the PPRK from above, sensor 2 is the next sensor clockwise (to the left of the controller), and sensor 3 is the remaining sensor.

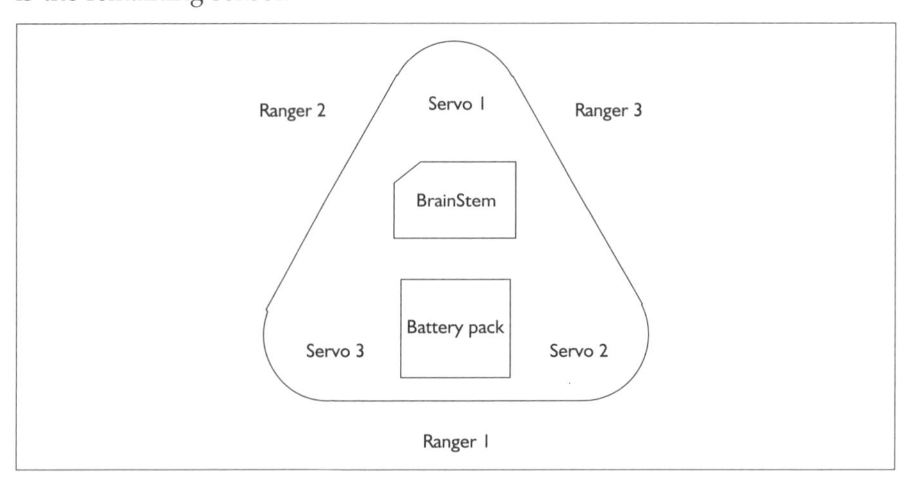

Checking the Digital Ports

The PPRK we built in Chapter 3 does not have anything attached to the digital ports of the BrainStem controller. If you did have any digital devices attached to those ports, you can check them with the GP application. Select the Digital tab now. You will see five rows of controls labeled Digital 0 to Digital 4.

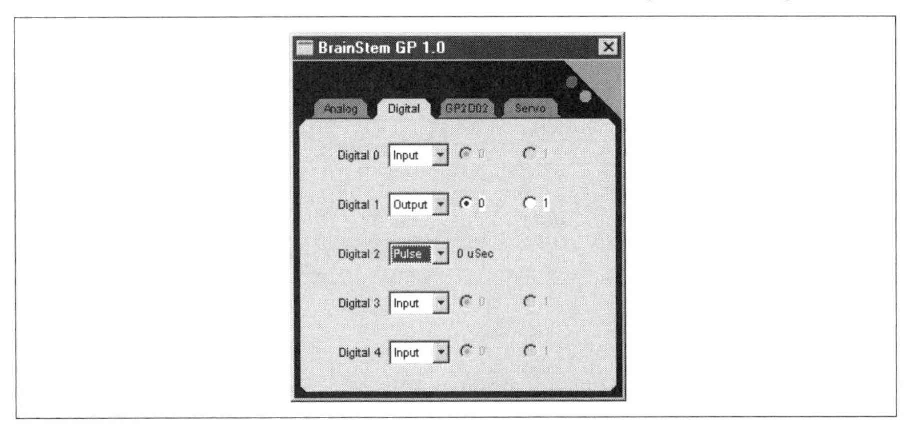

Select Input, Output, or Pulse using the drop-down combo box. With Input selected, the radio buttons will show what value a digital device is sending to the port. With Output selected, you can select one of the radio buttons to send a 0 or 1 value to the digital device attached to the port. The digital device attached to the port should respond to the value you send. The Pulse setting shows the rate that pulses are received from the port.

GP2D02

The GP2D02 is another type of Sharp sensor. Since the PPRK does not use this sensor, you will not make use of this tab of the GP application—you can safely ignore it.

Servo

This tab includes two important functions. First, it allows you to check the operation of the servos. Four horizontal sliders correspond to each of four possible servos. Since the PPRK uses only three servos, only three of the four sliders will be used. The slider starts with the control centered. By moving the slider to one side of center, the corresponding servo will rotate in one direction. By moving the slider to the other side, the servo should rotate in the opposite direction. You can also use this tab to check whether the servos are connected to the correct port. Recall that servo 0 is closest to the BrainStem controller (opposite sensor 0). Servo 2 is the next servo clockwise when looking at the PPRK from above, and servo 3 is the remaining servo.

The second use for this tab is to recenter your servos if they become uncentered. The servo should operate only when a program is commanding it to operate. If the servo rotates when you first turn on the robot and no program is running, the servo is no longer centered. Click the Config button to display the Servo Configuration dialog box shown here. By moving the Offset slider, you can adjust the servo so that it is centered.

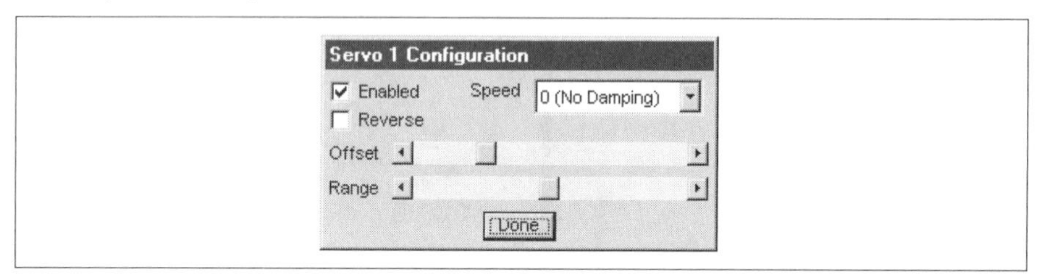

The Console

Unique versions of the Acroname Console program can be used for a number of different platforms. This program lets you compile, load, and launch programs for the BrainStem. The programs are written in a special language called TEA, which stands for Tiny Embedded Applications. If you want to write programs in C/C++, Java, or BASIC, you will not need the Console program. We will look at programming in those languages in Chapter 7.

To install the Console application to your Palm OS device, locate these files, which need to be loaded to the PDA:

- console.prc
- aIO.pdb
- aLeaf.pdb
- aSteep.pdb
- aSteepGen.pdb
- aSteepOpt.pdb
- aStem.pdb
- aTEAvm.pdb
- aUI.pdb

NOTE: Some of these .pdb files are used by other applications. If you've already installed one of the .pdb files for the Config or GP applications, you do not need to reinstall it. Also, if you download a later version of Console from Acroname, install all the .pdb files that come with it regardless of whether or not they are listed here.

After all the files have been loaded, start the Console application by tapping the Console icon in the Application Launcher. For the Windows or Mac version of the application, launch the Console executable from the \aBinary directory.

Like the GP application, launch Console first, and then turn on the robot. When the BrainStem connects to the application, you will see a heartbeat light on the controller and the Console application, shown next. One other thing to keep in mind when running the Console application on a PC: The Console application runs a mini web server. If you already have a web server running on your PC, the two programs might interfere with each other. If that happens, you will need to shut down the other web server temporarily while you use the Console application.

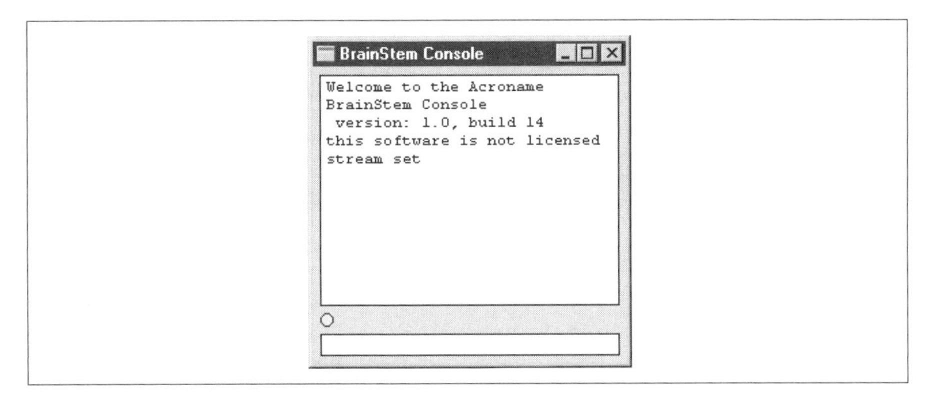

Sample TEA Programs for BrainStem

A number of sample programs are provided by Acroname for running the BrainStem PPRK. These sample programs are written using the TEA language. The Console program is used to compile, download, and launch TEA programs.

The Console program operates on files in the file system, so you don't have to worry about loading or opening any files. Also, if you keep the default directory structure used by Acroname, the Console program automatically knows where to find the files it needs. On a PC, it looks for TEA files in the \aUser and \aSystem directories; on a Palm OS device, it looks for TEA files as memos in the Palm Memo Pad.

Acroname provides four sample programs for you to try out with the BrainStem. The four programs are simple.tea, chase.tea, runAway.tea, and wallHug.tea. On a PC, they are located in the \aUser directory. For the Palm distribution, they are located in the \brainstem directory on your PC. TEA source code files use the filename extension *.tea*.

The Robot Geek Says

Fun Facts to Know and Tell About Programming

For the most part, we assume that if you want to program the robot, you already know how to write software—at least a little. If you are not a programmer, here are some fun facts to know about software programming:

- You usually write a program using human-readable words. The code you write is known as *source* code. Many different languages for writing code are available, including C, C++, BASIC, and Java.
- The computer doesn't understand human-readable words—at least not yet—so the program is converted into a machine-readable form, also known as *binary* code or *executable* code.
- The process of converting the source code to binary code is generally known as *compiling*.

Load TEA Files to the PDA If you are running the Console program on your PC, skip ahead to the "Compile a TEA File" section. Before you can compile any programs on your PDA, you need to load them to the PDA. TEA files are stored as memos on your PDA. The first line of the memo must contain only the name of the program. The easiest way to do this is to create the memo on your Palm Desktop (or in Outlook, if that's the program you use to synchronize with your PDA) and then perform a hot sync operation to load it to the PDA.

Find the simple.tea program in the \brainstem or \aSystem directory and copy it into a memo or an Outlook note. Add the line simple.tea as the first line of the memo:

```
simple.tea
/* simple.tea                                         */
/* BrainStem PPRK program                             */
/* Wait till back sensor is approached, then do a     */
/* little jig.                                        */
```

Note that even though the original first line (the second line in this listing) did contain the name, it was part of a comment (everything between /* and */ is a comment) and so would not be recognized as the program name.

If you look at the next few lines of the memo, you'll see some statements that look like this:

```
#include <aCore.tea>
#include <aPPRK.tea>
#include <aA2D.tea>
```

These are TEA files used by the program simple.tea. These files will also need to be added to the PDA as memos.

Find these files in the \brainstem or \aSystem directory and perform the same steps to create memos or notes of each of these files. aCore.tea and aA2D.tea also include aIOPorts.tea, so you will need to create a memo or note for that file. Here is a list of all the files you need for simple.tea:

- simple.tea
- aCore.tea
- aPPRK.tea
- aA2D.tea
- aIOPorts.tea

When you have created memos for all these files and loaded them to your Palm OS device, you are ready to compile the program.

Compile a TEA File You prepare a TEA source file for the BrainStem by compiling it. The command for compiling a program is steep. Enter this command

```
steep "simple.tea"
```

in the Console command line. Press ENTER to execute the command. For a PDA, place the cursor at the end of the line, and perform the Graffiti stroke for Enter (make a diagonal stroke from upper right to lower left in the Graffiti area).

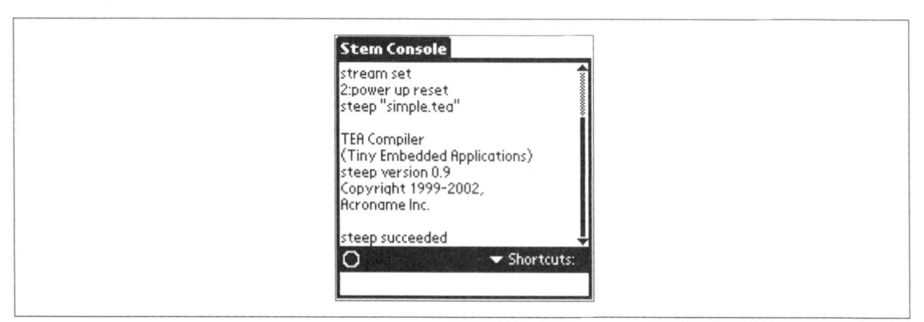

Assuming you created the memos correctly, or you kept the directory structure provided by Acroname for the PC, the Console application should find the TEA file and compile the source code into binary code. This binary code is stored as a new memo on your PDA or in a file in the \aObject directory of your PC, using the same filename, but with the extension *.cup*. Thus you now have a simple.cup memo on your PDA or a simple.cup file on your PC.

You are now ready to load the program to the BrainStem. If you are using a PDA, you can find simple.cup as a memo. The cup files on a PC are located in the \aObject directory. *Do not attempt to edit the cup files in any way!* They are binary files, so they won't make any sense to you anyway, and modifying them will certainly cause the program to fail.

Download a cup File to the BrainStem Here is the general form of the command to download a program from the Console application to the BrainStem:

```
load "file_name" module_id slot
```

The first parameter in the command is the quoted name of the compiled TEA file, including the .cup extension. The second parameter is the module number. For the BrainStem GP 1.0, *2* is the default value for the module number. Since we haven't shown you how to change it, we'll assume that *2* is the correct parameter for the module number. The final parameter is the *slot* number. The BrainStem module has 10 *slots* numbered from 0 to 9. This allows the BrainStem to store 10 programs at any one time. You can switch between the programs with the launch command that we will see in the Launching a CUP File section. So, here is the command to load simple.cup to slot 0 of the BrainStem:

```
load "simple.cup" 2 0
```

If the load command is successful, the Console will print a message that says the file was loaded.

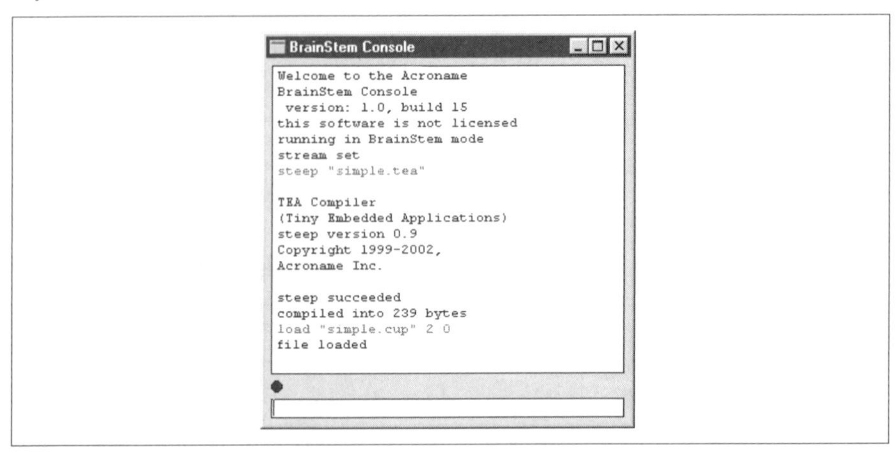

If there is a problem, you will see an error message.

Launch a cup file

After a cup file has been downloaded to the BrainsStem, you will send a launch command to start the program:

```
launch module_id slot
```

The first parameter in the command is the module number. Again, for the BrainStem GP 1.0, 2 is the default module number. The second parameter is the same slot number to which the program was loaded. The command to launch the program in slot 0 is shown here:

```
launch 2 0
```

And, the entire sequence of commands to compile, load, and launch the simple.tea program using slot 0 on the BrainStem module will look like this:

```
steep "simple.tea"
load "simple.cup" 2 0
launch 2 0
```

If everything works correctly, the program will start executing and the robot may begin to spin its wheels. If you are using a PDA, simply place the robot on the floor. If you are using a PC, detach the robot from the serial cable and place it on the floor. Try placing an obstacle in front of sensor 0 and see what the robot does.

Examining a TEA File

When we compiled, loaded, and launched the simple.tea program on the PPRK for the first time, it didn't work the way we expected it to. The documentation in the manual states

> *Wait until the robot stops moving and then wave your hand in front of each ranger slowly. One of the rangers will initiate a simple movement program....*

We interpreted that to mean that if *any* sensor detected an object, the robot would perform the movement program. But when we launched the program, two of the three sensors did not appear to be detecting anything, because when those two sensors should have detected something, the robot did nothing.

Next, we checked all the connections and used the GP and Config programs to check the sensors. Everything seemed to be okay, so why didn't it work the way we thought it should?

That's when we decided to look at the program itself to see what it was really supposed to do, rather than assuming it was supposed to work a certain way.

Let's take a quick look at that TEA file, simple.tea. We will cover the TEA language in more depth later in the book in Chapter 8, so we'll take just a quick sip of the program for now. The first clue to whack us right between the eyes was a comment line at the top of the file:

```
/* Wait till back sensor is approached, then do a little jig.  */
```

Obviously, the comment refers only to one sensor. Perhaps the robot was working exactly the way the program told it to work.

As an aside, it's worth noting that the comment line uses C-style comments. As we look at this program further, you will see other signs that the TEA language syntax is partly based on the C language. Another similarity to C is the ability to include other source files in your TEA file and use the functions defined in those files. The source file simple.tea includes three other source files. Here is one of the #include lines from simple.tea:

```
#include <aCore.tea>
```

The aCore.tea file has methods for pausing the robot and reading and writing the BrainStem ports. The software from Acroname includes several other TEA files in the \aSystem directory. These files contain lots of different functions that you can call from your TEA code, which simplifies developing new code.

Within your TEA source files, you can define functions that can be called. The simple.tea programs includes two functions: jig( ) and main( ).

The Robot Geek Says

The Rain in Spain Stays Plainly in the main( )

- The main( ) function is a special function that is called by the system code to start the program.
- C, C++, and Java also start programs with a main( ) function (although in Java, it is referred to as the main( ) method).
- The main( ) function can appear anywhere inside the source code.

So, back to our little story. We guessed that the jig( ) function was used to move and turn the PPRK when a sensor detected an object. If you look at the jig( ) function, you can see that it uses several functions from the other TEA files, namely **aPPRK_Go( )** and **aCore_Sleep( )**, to do its work:

```
void jig()
{
  aPPRK_Go(0, -100, 100, 0);
  aCore_Sleep(20000);
  aPPRK_Go(0, 0, 0, 100);
  aCore_Sleep(30000);
  aPPRK_Go(0, -100, 100, 0);
  aCore_Sleep(20000);
  aPPRK_Go(0, 0, 0, -100);
  aCore_Sleep(30000);
  aPPRK_Go(0, 0, 0, 0);
}
```

The other function in the source code is the main( ) function. This is where we found the answer. If you look at the following source code, you'll see only one line of code that appears to read a sensor, and it appears to read IR1. That would explain why only one sensor caused the robot to do its little jig:

```
void main()
{
  int b;
  int range;

  /* five second wait before starting              */
  aCore_Sleep(25000);
  aCore_Sleep(25000);

  /* loop continuously                              */
  while (1) {
    /* take a reading                               */
    range = aA2D_ReadInt(APPRK_IR1);
    /* see if something was present and do         */
    /* the jig if so                                */
    if (range > 100)
      jig();
    /* stall for 1 second                           */
    aCore_Sleep(10000);
  } /* while */
} /* main */
```

Now What?

Now that you've had the chance to try some sample programs with your PPRK, you're probably ready to write some programs of your own.

In Chapter 6, we introduce the Palm Robot Programmer, an application written specifically for this book. You can use the Palm Robot Programmer to create PPRK programs in a visual style by adding icons onto a workspace.

If you are already a programmer and want to get your hands on some code, you should check out Chapters 7 and 8. In those chapters, we show you how to use various tools for writing programs for the PPRK. You've already seen some of those tools in this chapter. We'll look at those and other tools and get into some more in-depth programming for the original PPRK using the Pontech SV203 controller in Chapter 7. In Chapter 8, we'll look at tools for programming the BrainStem PPRK. See you there!

Chapter 6

The Palm Robot Programmer (PRP)

In the beginning, scientists programmed their computers by flipping switches. The position of the switch corresponded to a bit, either 0 or 1, and the computer stored each bit. After entering all the bits that made up the program, the scientist told the computer to execute the program. One of the first computers, called the ENIAC, could execute 5000 operations per second and used 1000 square feet of floor space. Scientists predicted that soon computers would be able to program themselves.

Next came punch cards and Teletype terminals. Engineers could enter programs using simple languages that were converted into machine language and executed by the computer. Computers were built using solid-state circuits rather than vacuum tubes, and thus they began to shrink in size. Soon, they said, computers would be able to program themselves.

Not long afterward, more powerful languages were developed, such as Algol, FORTRAN, Smalltalk, Lisp, and C—to name a few. Engineers had remote terminals, while the software ran on a mainframe computer. Gone were the punch cards and batch runs of the old days. A program could be entered at any time, and the engineer would be able to compile and run the program immediately. Scientists were using the new languages to create expert systems and experiment with artificial intelligence. We were finally within reach of creating computers that could think and program themselves.

The PC revolution began in the 1970s with computers made by companies such as Commodore and Apple. At this point, anyone could learn to program using BASIC, the Beginner's All-purpose Symbolic Instruction Code. In the 30 years since, computers have become more powerful; the ENIAC computer can now fit on a chip barely 1/4-inch per side, and handheld computers like the Palm offer dramatically more processing power than those early room-sized computers. New languages, such as C++ and Java, have entered the scene. And scientists continue to predict that computers will soon be able to program themselves. Don't we wish it were so.

Unfortunately, neither your computer, nor your robot, can program itself. As you saw in the last chapter, your robot needs a program to tell it how to move and react. The programs in the previous chapter offer a starting point that lets you immediately see your robot in action. But the real fun comes when you are able to tell your robot how to behave by writing your own program.

Don't get scared off if you aren't a programmer.

As we've hinted at several times in the book, we are going to make it a little easier for you to write programs if you're not a programmer. We've developed a software application that we've named the Palm Robot Programmer, or PRP for short, which does two important things for you: it makes programming your robot easy and flexible.

Using the PRP is easy because all you need to do is click icons to create a program visually (Figure 6-1). It's also flexible because you can use the PRP to create programs in several different languages. This chapter will show you how to use the Palm Robot Programmer to create your own Palm Robot programs.

Figure 6-1
The PRP allows you to create programs using icons that represent parts of the program.

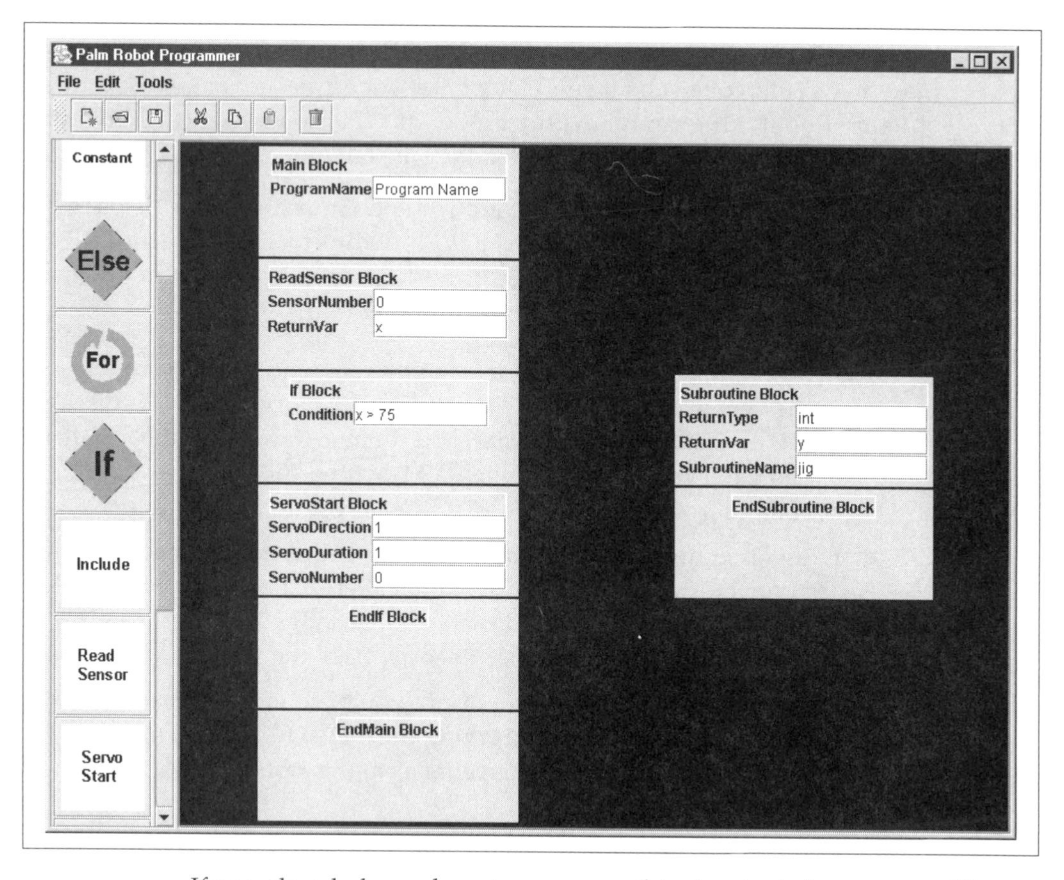

If you already know how to program, this chapter is for you, too. We provide the PRP tool to make it easier for nonprogrammers to program their robots, but nothing stops you experts from using this tool as well. And if you're a real hacker, and you want to write the code yourself, we offer two things: The PRP can create source code in several different languages. At the end of the chapter we'll show you how to create a plug-in to the PRP so that it can create source code in any language of your choosing. Second, later chapters in this book show you how to use your programming skills to create programs for the Palm Pilot Robot Kit (PPRK) using Basic, C, TEA, and Java.

Installing the Palm Robot Programmer

We think you'll find the installation process pretty painless. The installer is located on the CD-ROM in the folder\Palm Robot Programmer. If you're using Windows, you'll find the installer in \Palm Robot Programmer\Windows\VM\ or \Palm Robot Programmer\Windows\NOVM\; the installer is named PRP_install.exe. For Mac OS X users, look for \Palm Robot Programmer\MacOSX; the installer is named PRP_install.zip. Last, but not least, we haven't forgotten about you Linux folks. Your installer is in the directory\Palm Robot Programmer\Linux\VM or \Palm Robot Programmer\Linux\NOVM; the installer is called PRP_Install.bin.

You may have noticed that the Windows and Linux directories listed in the previous paragraph have a VM and NOVM subdirectory. Because the PRP application is written using Java, you must have Java Runtime Environment version 1.4 or later installed on your computer.

- If you don't already have a Java environment, you'll want to use the version of the installer in the VM directory of the CD-ROM. VM stands for *virtual machine*, and it refers to whether the installer has the Java Virtual Machine (JVM) with it (not whether your computer has the JVM).
- If you know that you *do* have the Java 1.4 runtime already on your computer, you can use the installer that does *not* come with its own VM; you will find that installer in the NOVM directory. (NOVM means the installer does *not* have the JVM.)
- If you don't know whether or not you have Java on your computer, we recommend trying the NOVM installer first; most computers already have Java, so this option will work for most people.
- If you use the NOVM installer and the program doesn't work, you'll need to use the VM installer.
- If you're a Mac OS X user, you can get a Java 1.4 version from www.apple.com/java.

By and large, installing PRP is essentially the same whether you are using Windows, a Mac, or a Unix/Linux system. If you're not using Windows, the main difference in the installation process will be in the directory structure used on each system and the look and feel of the actual dialog boxes.

1. Start by copying the installation file to your hard drive. (Yes, you can run the installer from the CD-ROM, but it runs more efficiently from the hard drive.)

TIP: We suggest creating a special directory on your hard drive named "Install," in which you can save the installation programs for various applications you've downloaded. Kevin's PC, for instance, has a directory for the PRP installer called C:\Install\PRP. This makes it easier to find the install files in the future if you ever need them again.

2. Double-click the installation program to launch it. You should see the Install Wizard's first screen, as shown in Figure 6-2.

Figure 6-2
The introduction screen of the installer provides an overview of the installation process.

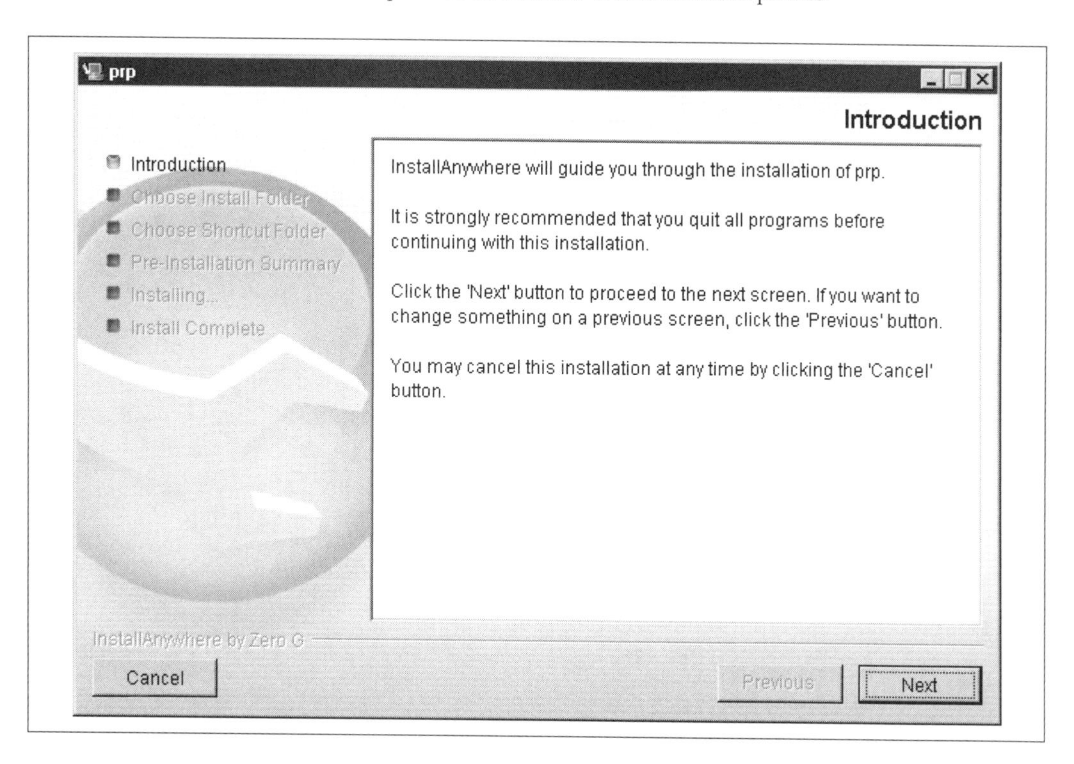

3. Click the Next button. The installer then shows a screen that asks you where you want to install the program, as shown in Figure 6-3.

Figure 6-3
This screen allows you to choose the installation folder for the PRP.

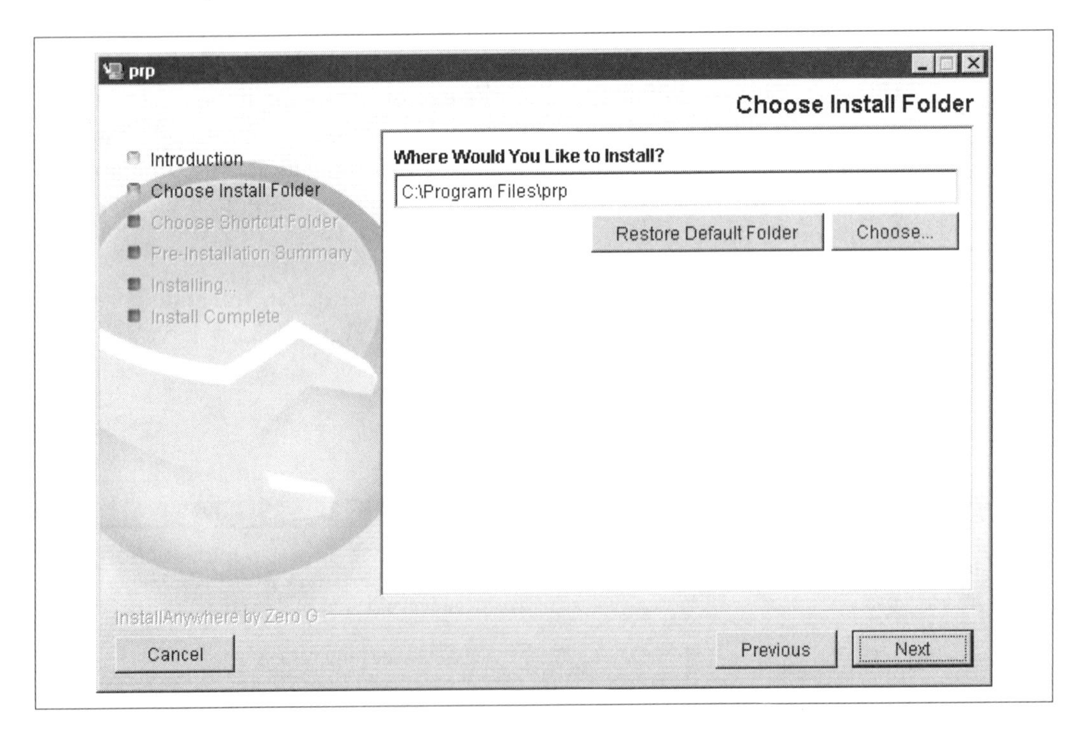

4. For Windows, the default folder is C:\Program Files\prp. If you're using a Mac, it'll be the /prp folder in the normal programs directory. Unix/Linux users, you'll see /prp in your normal programs directory. Want to select a different folder? No problem—you won't even hurt our feelings.
5. After selecting the appropriate directory, click Next. The Installer now asks you where to create program icons, as shown in Figure 6-4.

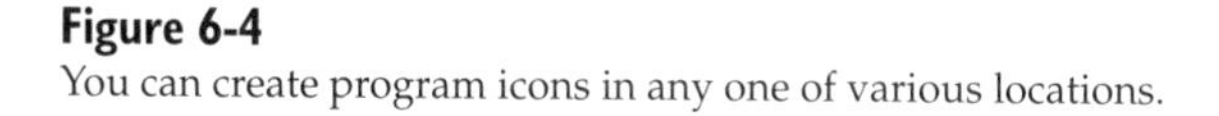

Figure 6-4
You can create program icons in any one of various locations.

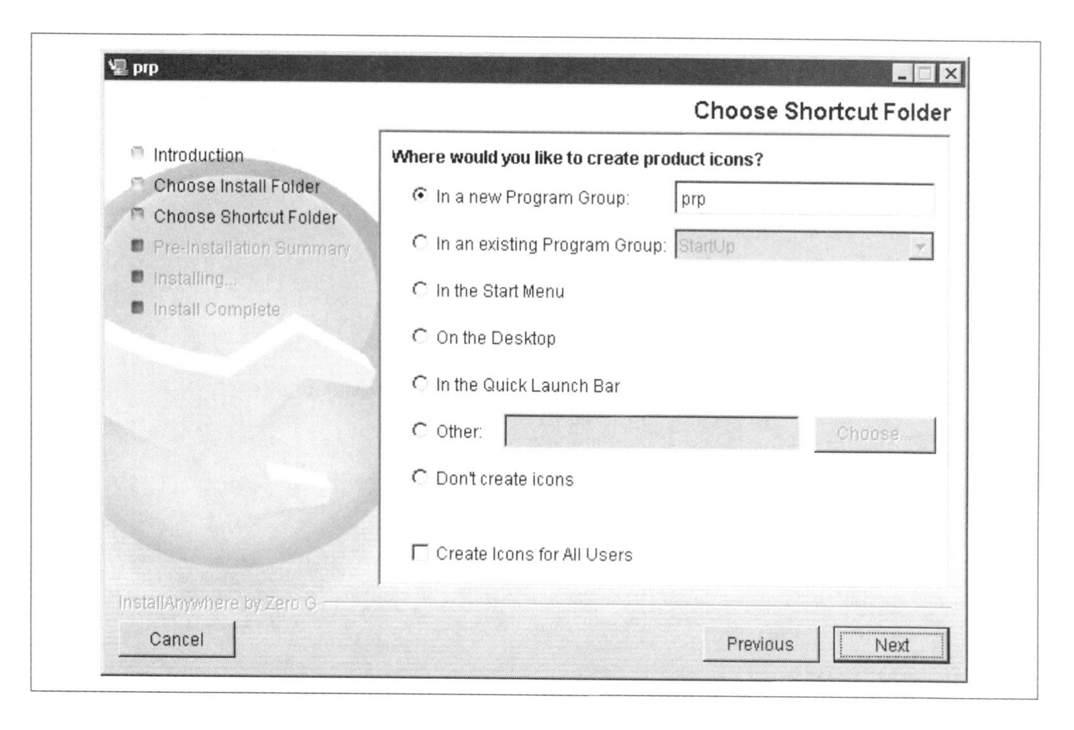

6. This will create program shortcuts in the given location. Because these options are offered as radio buttons, you can choose only a single location. However, if you really want program icons in several locations, it is easy to copy the shortcuts after the installation is complete.

7. Click Next to see a summary of your installation options (which you can see in Figure 6-5). Of course, you can go back at any time and change any of these options by clicking the Previous button.

Figure 6-5
The Installer presents a summary of the installation options before completing the installation.

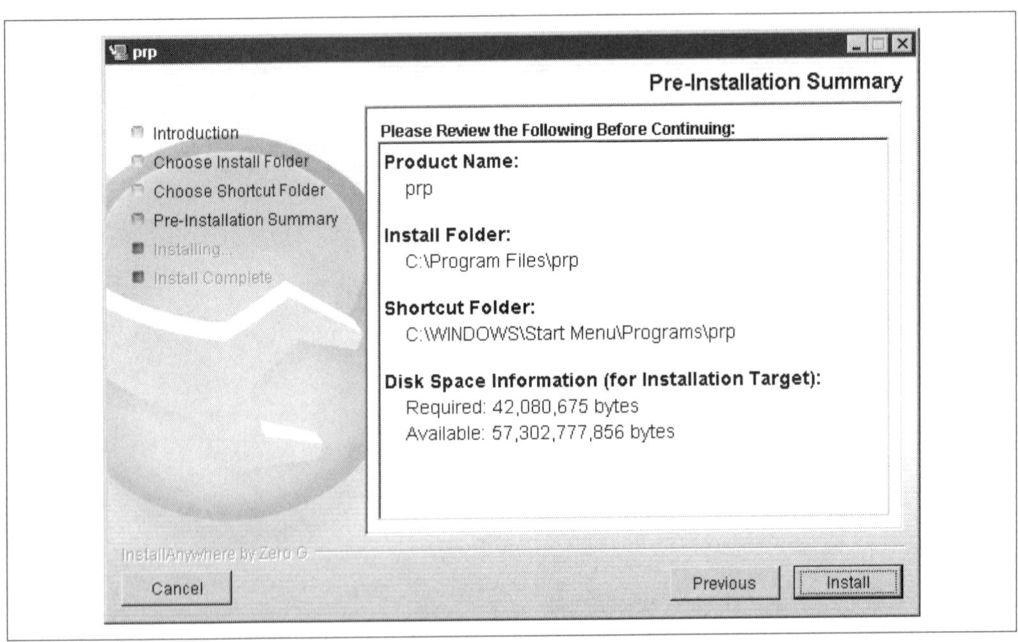

8. If you are satisfied with your choices, click the Install button and the installer will install the PRP application to your computer. When the installation is complete, click Done.

Using the PRP

Now that we've installed the PRP, let's create our first program. We'll whip up something simple so you can see how all this works before we bore you … er, that is, before we review the program in loving detail.

NOTE: This is for the techie sticklers out there. We've already referred to the "programs" you create with the PRP, and we will continue to do so in this chapter. To be accurate, the PRP creates only a *representation* of a program. When programmers write a program, this usually means they type source code into an editor. The programs you create with the PRP are not source code; they're an abstraction of source code that is later translated into source code. However, it's a little awkward to write, "the program representation you create with the PRP," so we'll continue to refer to what you create with the PRP as a program. If you want to see what the PRP program looks like before it gets translated into source code, we'll show you that later in the chapter in the section "How the PRP Converts a Program to Source Language."

The PRP Window

Find the program icon that was created by the installer and launch the program. Alternatively, if you are using a command line, navigate to the installation directory and type the program name on the command line to launch it. When the program launches, you will see the PRP window.

The PRP window contains three main areas: the menu bar and toolbar, the component palette, and the workspace (as shown in Figure 6-6). The menu bar and toolbar are at the top of the window, similar to their placement in most windowed programs. The component palette runs along the left side of the window and contains the various program components you can use to build programs. Finally, the biggest pane in the window is the workspace. You will add program components to the workspace to build a program.

Figure 6-6
The PRP window: components are added from the palette to the workspace to create programs, while major program functions are accessed through the menus or toolbar.

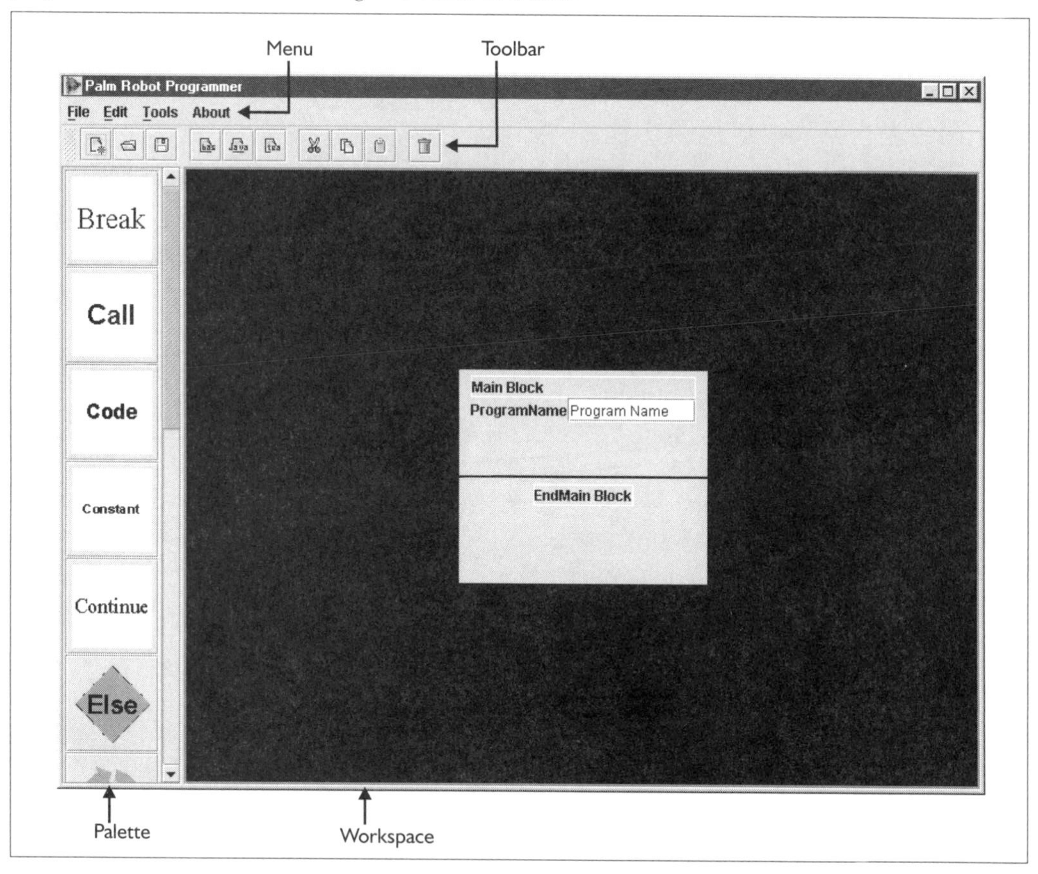

Making Your First Program

Ready? Let's get started!

1. When you start a new program, the PRP automatically adds one block to the program. This is the Main block, as seen in the following illustration. The Main block has a corresponding EndMain block. These two blocks mark the start and end of your program. The Main component provides a means for the PRP to know where your program starts and stops; it also allows the PRP to differentiate the core of the program from subroutines (which we will look at soon, so stop getting so antsy).

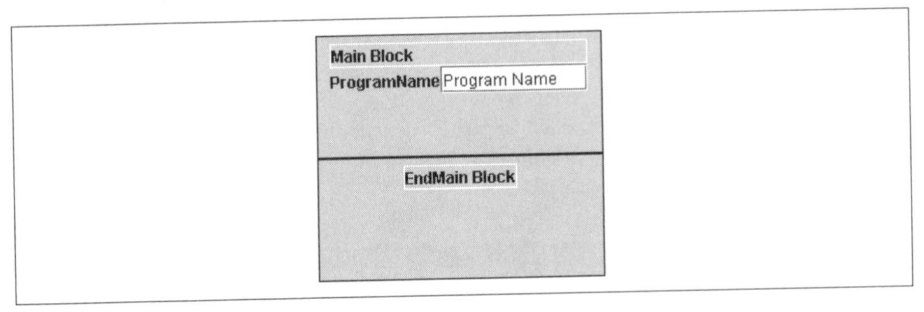

2. In the palette, find the component named Servo Start and click it.
3. Position the mouse over the Main block in the workspace and click the mouse. The PRP will add the Servo Start component to the Main component and draw the ServoStart block in the workspace. Notice that you don't drag the component onto the workspace as you might do in other drawing programs. You simply click the component in the palette and then click a component in the workspace.

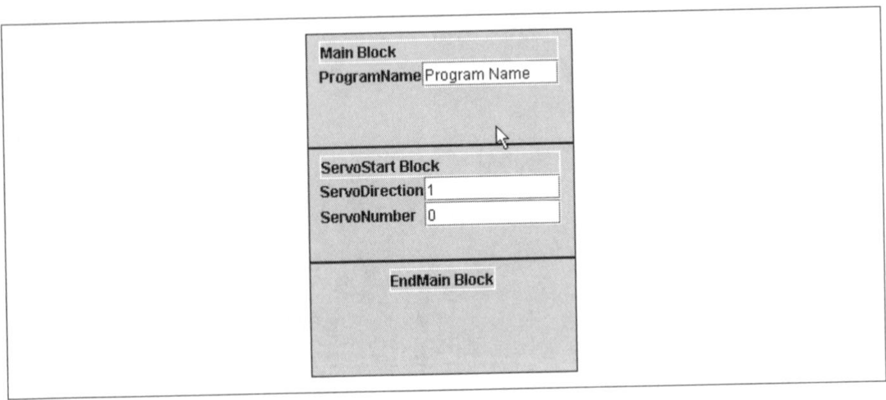

4. Repeat this three more times, so that you have a total of four Servo Start components in the workspace. Notice that the program is drawn visually from top to bottom of the workspace.

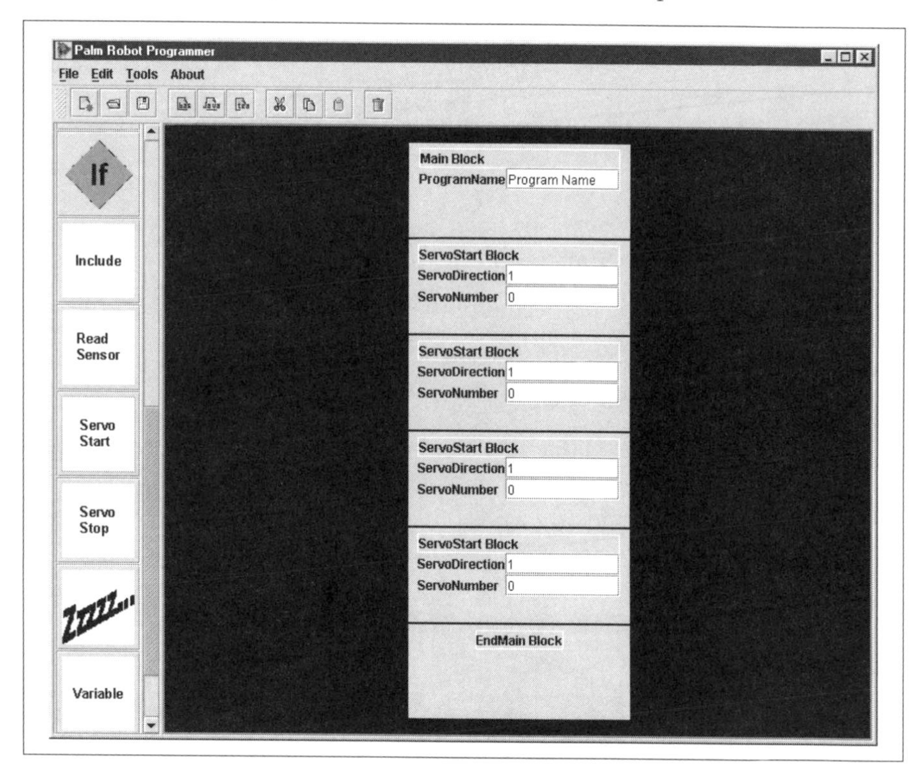

In the next few steps, we'll refer to component one, component two, and so on. We're actually referring to the *order* of the components starting from the top of the program—*not* the order in which the components were placed onto the workspace. You are able to add components to any position within the program, so the one-hundredth component to be added could easily be the topmost, or first, component in the program

5. Displayed within each block are the properties for the block. The properties allow you to change the way a particular program component works. For the ServoStart block, we can set two properties: the ServoNumber, and the ServoDirection. For the PPRK, the ServoNumber will be 0, 1, or 2. The ServoDirection should be 1 for forward or –1 for reverse.

6. Set the ServoDirection to 1 (duration can be either 1 for forward or --1 for backward), and set the ServoNumber to 1. This will cause servo one to rotate in the forward direction.

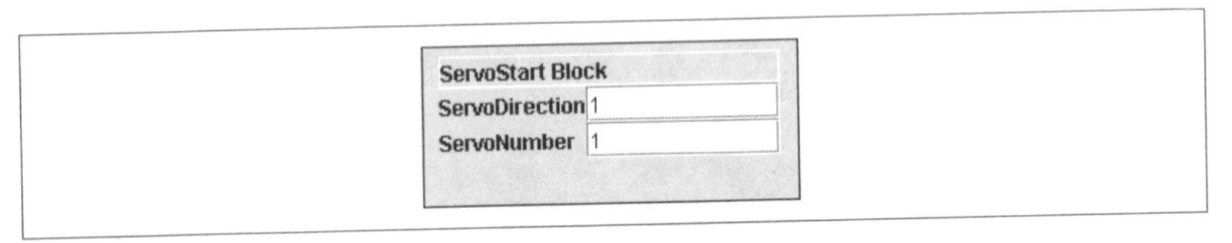

7. In the next ServoStart Block, set the ServoDirection to –1, and the ServoNumber to 2.
8. Set the properties of the third and fourth components as follows:
   - ServoDirection –1, ServoNumber 1
   - ServoDirection 1, ServoNumber 2
9. Now look in the palette for the component that has the picture of a string of Zs. This is the Sleep, or pause program, block. Click on this icon and add a Sleep block after the second ServoStart Block; add another Sleep block after the last ServoStart Block.

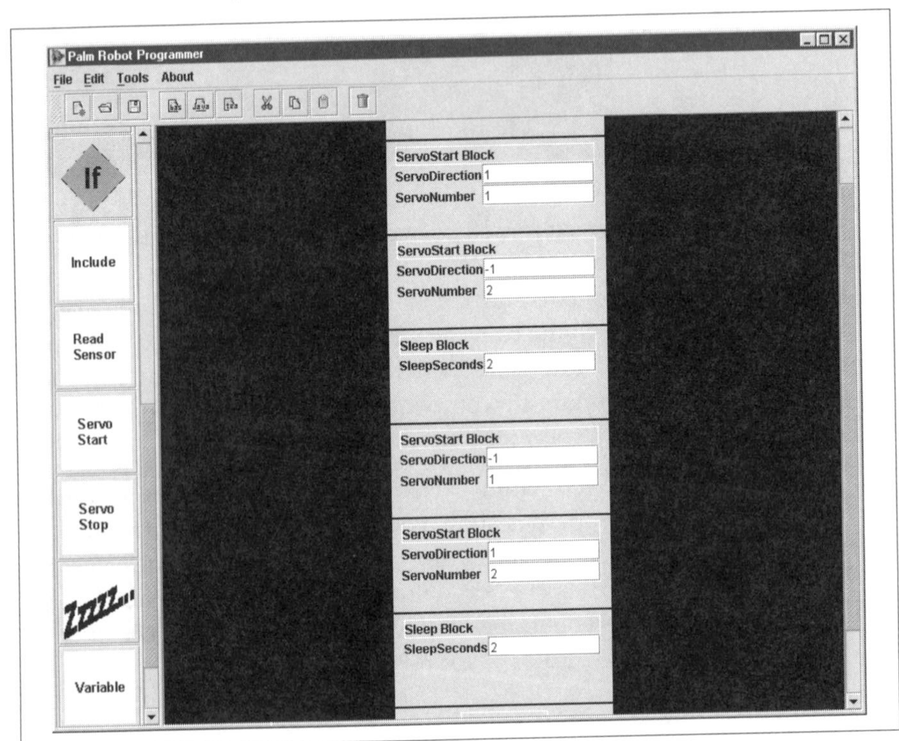

10. Set the SleepSeconds property of each block to 2.
11. Finally, add two ServoStop components. To do that, click the ServoStop component in the palette, position the cursor over the last Sleep component in the program, and click the mouse. Do this again to add a second ServoStop component.

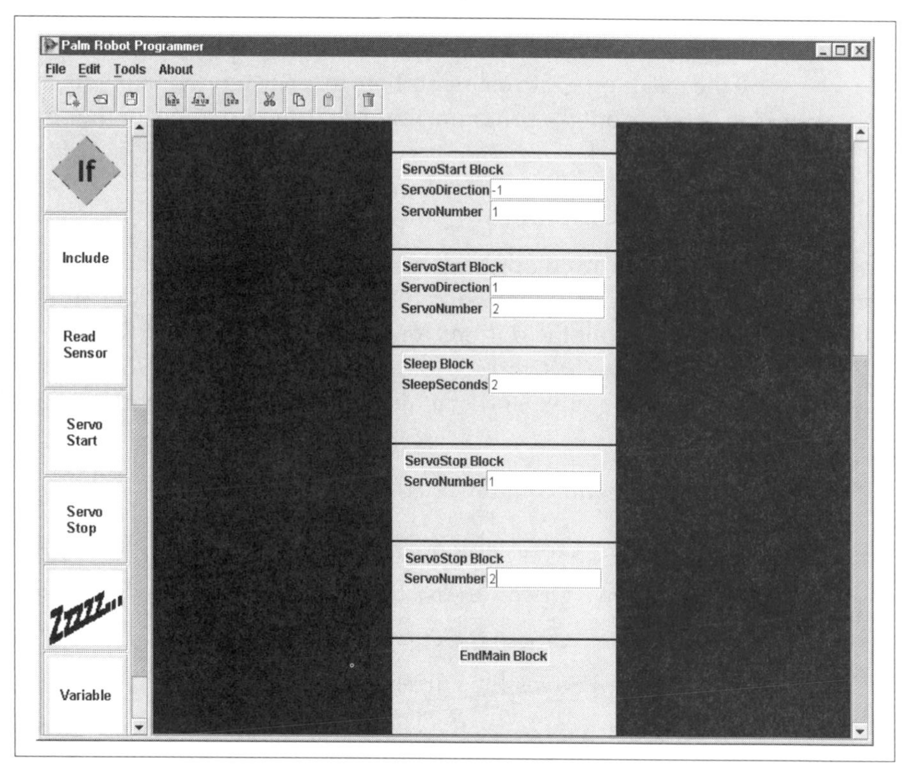

12. Set the servo number for one of the ServoStop Blocks to 1 and for the other ServoStop Block to 2.
13. At this point, it's a good idea to save your program. Choose File | Save and save the program to your hard drive.
14. Choose Tools | Convert. If you have the SV203 PPRK, convert the program to Basic and store the source code any place you want. If you have the BrainStem PPRK, convert the program to TEA and store the source code in the \aUser directory of your Acroname software.

The last step is to transfer the source code to the Palm or to the BrainStem controller and watch as your robot executes a fancy maneuver: it moves forward for two seconds, and then it moves backward for two seconds. We'll show you how to transfer your code later in the chapter, in the section "Loading Source Code to the Host."

The PRP in Detail

Wasn't that easy? In fact, Kevin is pretty sure that even Dave can handle that particular program. You probably don't need a detailed explanation of each button, menu option, and program component. Nonetheless, we've outlined all the major program elements here for easy reference. If you get snagged trying to figure out the program, this is the place to come for the straight scoop on how it all works.

Menus and Toolbars

Like all windowed applications, some of your interaction with the PRP will occur through the menus or toolbar. At the top of the PRP window is the menu that contains all the options for interacting with the PRP. Below the menu is the toolbar that has buttons for many of the functions you will most often use. The menu is simple and contains only a few major entries: File, Edit, Tools, and About.

The File menu contains the following options:

- **New** Closes any open program and starts a new program
- **Open** Opens an existing program from the file system
- **Close** Closes the current program
- **Save** Saves the current program
- **Save As** Saves the current program with a new name
- **Exit** Exits the PRP

The Edit menu contains the following options (there are no surprises here, guys ...):

- **Cut** Deletes the currently selected component from the workspace, and the deleted component is saved in a copy buffer
- **Copy** Copies the selected component into a copy buffer
- **Paste** Pastes the contents of the copy buffer onto the workspace following the currently selected component
- **Delete** Deletes the currently selected component from the workspace, and the deleted component is *not* saved in a copy buffer
- **Add Subroutine** Adds a Subroutine block to the program

- **Delete Subroutine** Deletes a Subroutine block, and all the blocks within the Subroutine from the program

The Tools menu contains the Convert To option:

- **Convert To** Converts a PRP program into source code. There are three submenu choices for Convert: Basic, Java, and TEA. Each option translates the program to the named language.

The About menu contains the single item, About, which displays the version number of the program. The toolbar contains quick-access buttons for many of the menu options you've just seen.

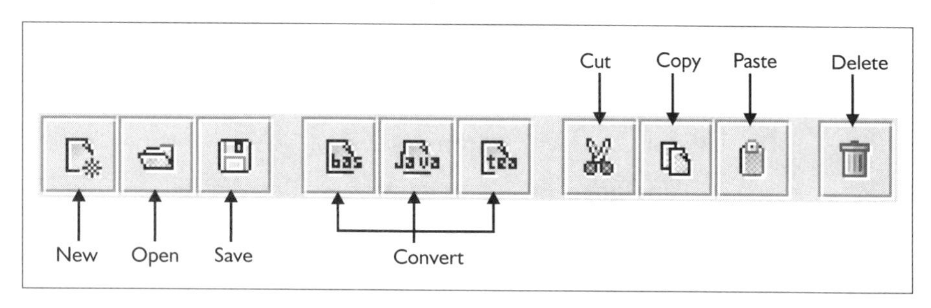

The Component Palette

On the left side of the main window is the component palette. This contains icons for all the programming components you can use within your programs. It includes components for loops, logical decisions, variables, subroutines, and a few components that are uniquely specific to the PPRK.

The Variable Component This block allows you to create variables in your program. Variables are like aliases for values. They represent values and can be used to store values that can be used later. This component has a property for variable name, variable type (more on that coming up) and initial value (which is an optional property). For example, if we take a reading from a sensor, we would store the value in a variable. Later we can access that particular reading by using the variable in the code.

Some languages use the variable type property to identify what data is stored in a variable. For example, in the Java programming language, a variable that stores a number cannot be used to store a string of characters. Although the PRP does not restrict what you enter into the Type box, if you enter a type that is not supported by the language, the program you create will not work.

Variable Block	
Value	value
Type	null
VariableName	x

If you are using the PRP to write a TEA program, you can use one of five kinds of variables: char, unsigned char, int, unsigned int, and string. Here's what they mean:

- **char** Any value from –128 to 127
- **unsigned char** Any value from 0 to 255
- **int** Any value from –32768 to 32767
- **unsigned int** Any value from 0 to 65535
- **string** A fixed series of characters

If you are writing a program for Java, you can use any legal Java data types. However, we recommend you limit yourself to the following:

- **char** Any value from –128 to 127
- **int** Any value from –32768 to 32767 (Java ints can actually have larger magnitudes, but to ensure compatibility with the BrainStem, you should limit ints to the values shown)
- **String** A fixed series of characters

Finally, limit your Basic programs to these three data types:

- **char** Any value from –128 to 127
- **int** Any value from –32768 to 32767
- **String** A fixed series of characters

The Constant Component Constants are like variables, except that they usually do not change within a program. Properties for this component are name, type, and value, and these properties have the same meaning as they do for variables. Why would you use a constant? Well, imagine that you are writing a program that reads data from Sensor 0. All through your code you use the number *0*. You know that this number is being used to refer to Sensor 0. The problem is that someone reading your program doesn't know what that 0 means. However, if you create a constant named Servo0 and set its value to 0, you can use the constant Servo0 in the code every time you want to use the value 0. Anyone reading your program has a better chance of understanding what Servo0 refers to in comparison to just 0. Java and TEA support constants. These languages will not allow you to change the value of a constant.

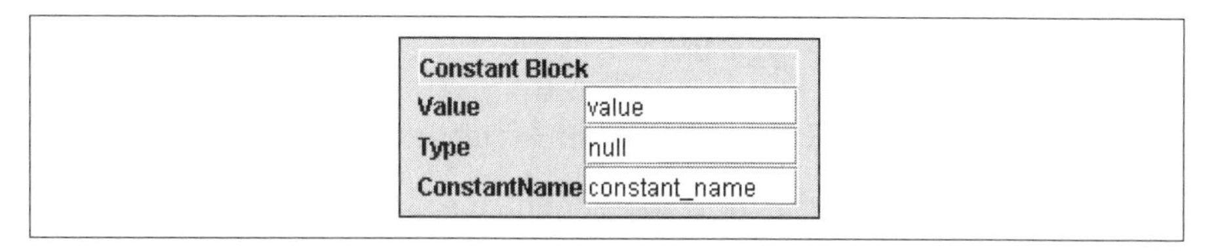

The If Component The If component allows your code to make a decision based on some condition. The property for this component is the logical condition to test for. When you add an If block to your program, the PRP automatically adds an End If block. Program blocks between the If and End If blocks are executed when the condition is true. The condition can use values, variables, constants, and any number of operators.

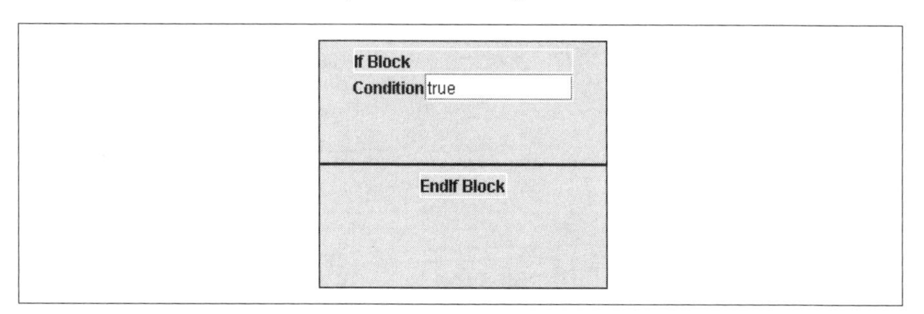

Some of the operators are listed here:

- == equals (TEA or Java)
- = equals (Basic)
- != not equals
- < less than
- <= less than or equal to
- > greater than
- >= greater than or equal to
- && and
- || or

Want an example? If you have a variable (let's call it x), you can test it for various conditions. The program might need to make a decision based on criteria: if x is greater than some value (x > 5), less than some value (x < 10), or equal to some value (x == 13), for instance. Using the **and** or **or** operators, you can further test for multiple conditions. For example, you can test if x lies between two values (the condition **x > 5 && x < 10** is read as "x greater than 5 and x less than 10." It is true if x is between 5 and 10); or whether x meets any one of several conditions (the condition **x == 2 || x == 4** is read as "x equals 2 or x equals 4" and means the condition is true if either x equals 2 or x equals 4).

In addition to the If component, the Else component is used when you want some code to execute when the condition is false. The Else block must be added somewhere between the If and EndIf blocks. When you add an Else block, the PRP automatically adds an EndElse block. Every program component between Else and EndElse will be executed if the condition property of the If block is false.

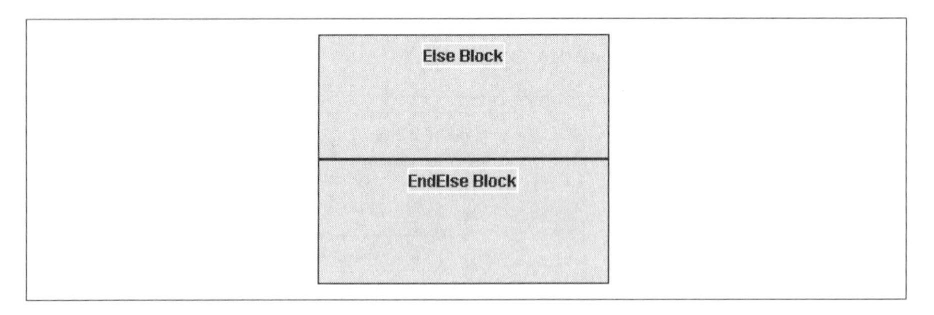

NOTE: The PRP will not stop you from adding an Else block anywhere within a program. However, if you place an Else block where it is not between an If and End If block, your program will not work.

The While Component A While component allows you to perform a loop until some condition is met. The single property for the While component is the *condition*. This condition meets the same parameters as the condition for the If component. When you add a While block to your program, the PRP automatically adds an EndWhile block. You can add components between the While and EndWhile blocks. When the While component is encountered, the condition is tested. If the condition is true, the components between the While and EndWhile blocks are executed. The program then goes back to the start of the While component. If the condition is false, the program goes to the next component after the EndWhile component; if true, it executes the components contained in the While component again.

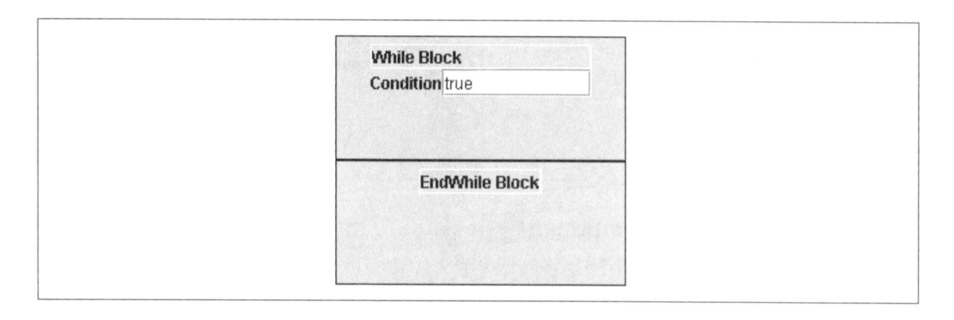

The For Component A For component allows you to perform a loop some set number of times. The For block contains a counter that is set by *properties*. The properties are the start value, the number of loops, and the step size. The step size is optional; if you don't specify one, it is assumed to be 1. When you add a For block to the program, the PRP automatically adds an EndFor block. You can add components to the program between the For and EndFor blocks. When the program encounters the For block, the counter is set to the start value. Then the program components between For and EndFor are executed. When the EndFor is reached, the counter is increased by the step size. If the number of loops has been reached, the program jumps to the next block after EndFor; otherwise, it goes back to the start of the loop and executes the blocks again.

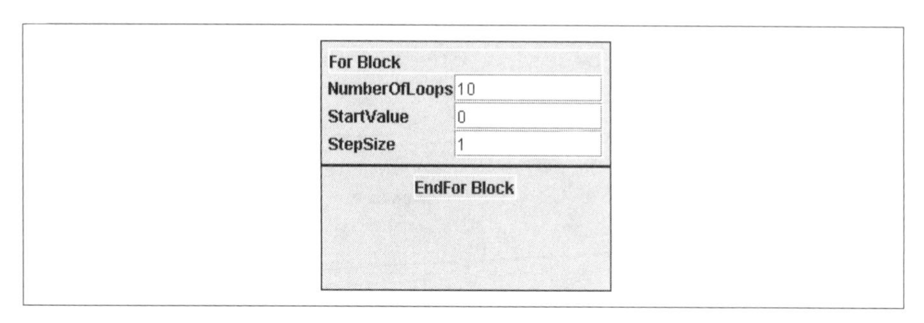

NOTE: Currently, the PRP supports only counting up. For example, if you set the start value to 1 and the number of loops to 10, the counter will count from 1 to 10 and then stop. However, if you try to set the start value to 10 and the step size to –1 in an attempt to count 10, 9, 8, ..., the program will not work.

The Continue Component If you are inside a While or For component, Continue will stop executing the loop immediately and jump back to the beginning of the For or While component. This component has no properties.

The Break Component If you are inside a While or For component, Break will stop executing the loop immediately and jump to the next component following the For or While component. This component has no properties.

The Call Component Call allows you to call subroutines by name. After the subroutine completes, control of the program passes to whatever component comes after the Call. The properties of this component are the name of the subroutine and a variable in which you can store any return value. If there is a return value, you also need to add a Variable block to declare the variable. If the subroutine does not return a value, the Variable field should be set to empty. Also, you need to ensure if the Subroutine does return a value, that the variable has the same type as the return value. We show you how to create a subroutine in the section "Creating a Subroutine, Function, or Method," later in this chapter.

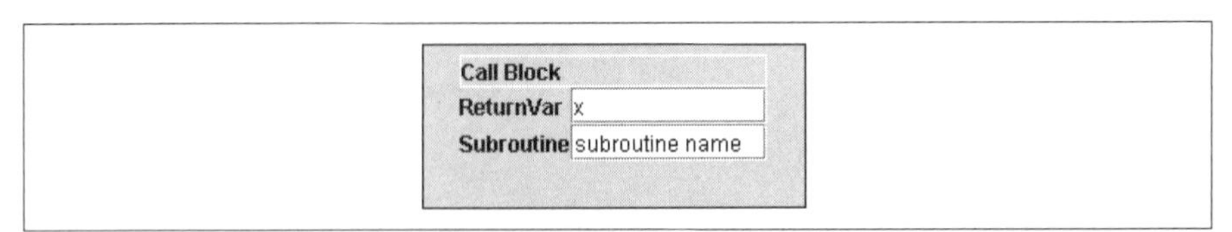

The Include Component This block allows you to include another source file into the program. This is provided for TEA and Java programs. TEA allows you to include other source files into a TEA program and call the subroutines in the other source file by name. The property for this component is the name of the source file. For example, if you install the BrainStem software, you will have a directory named \aSystem. This directory contains source code files that you can include. One such source file is aPPRK.tea. If you add an Include block and set the filename property to aPPRK.tea, the source file created for your program will have a statement to include the aPPRK.tea file and you can use aPPRK.tea functions in your program.

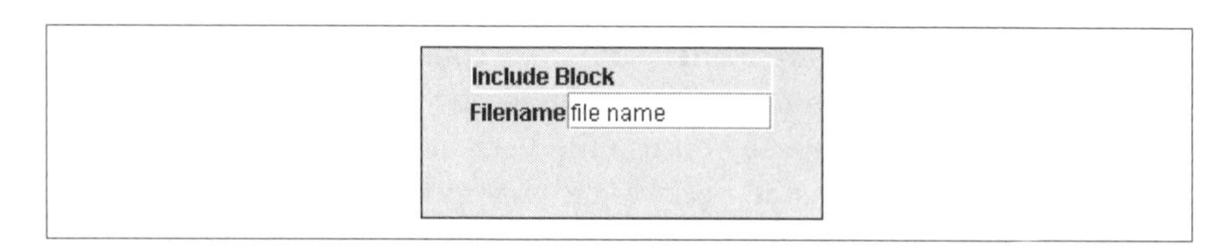

NOTE: If you convert your program to TEA, the PRP, will automatically include three source files for you: aCore.tea, aA2D.tea, and aPPRK.tea. You need to use Include only for other BrainStem TEA files.

Within Java programs, the Include block is used to tell the Java compiler which other classes your program uses. Classes are the binary form of Java code. You tell the compiler which other classes you are using in an import statement. The Include block is translated into an import statement in Java source files.

The Servo Start Component This component starts a given servo. Using *properties*, you can set the servo number from 0 to 4, the direction of the servo between forward (1) and backward (–1).

ServoStart Block
ServoDirection 1
ServoNumber 0

The Servo Stop Component This component stops a given servo. The servo number is set with a property and can range from 0 to 4.

ServoStop Block
ServoNumber 0

The Read Sensor Component This component reads a sensor. The properties are the sensor number (0 to 4) and the variable to store the sensor reading into.

ReadSensor Block
SensorNumber 0
ReturnVar x

The Code Component The Code component is used when you need to add a line of code that is not provided by any of the existing code blocks. For example, suppose you use the Variable block to create a variable named *x*. Later in your code, you want to change the value of *x* by adding some number to it or just setting its value directly. You could do that with the Code block. The property for the Code block is the line of code that you want in your program at that point. So, with the example just stated, you could change the value of *x* with a line of code like this:

```
x = x + 1
```

This line of code takes the value of *x*, adds 1 to it, and finally assigns the new value to the variable *x*. If you add a Code block and set its property to that line of code, that line of code will appear in the translated source file.

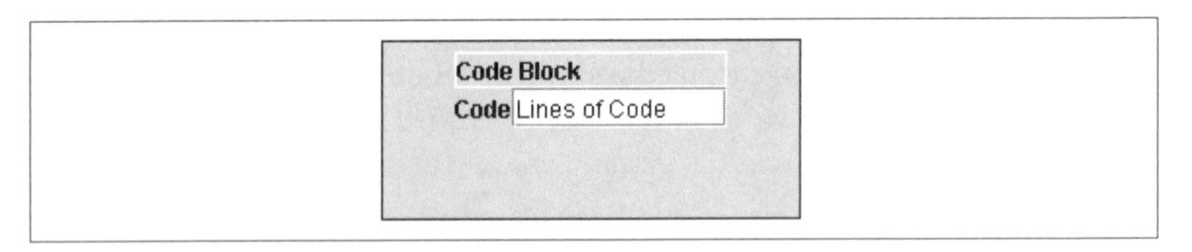

The Sleep Component The button for the Sleep block is labeled with a string of Zzzzzs. The Sleep block is used to pause the program for some number of seconds. After the number of seconds has elapsed, the program continues executing. The property for this block is the number of seconds to sleep.

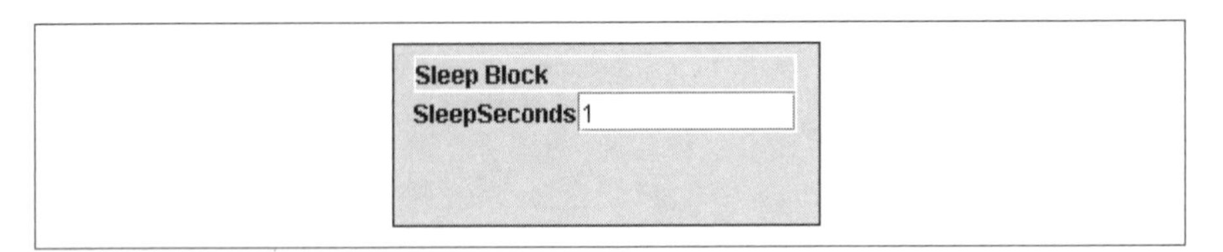

NOTE: If you are translating a program into TEA, you must be careful to set the seconds to a value no greater than 3.2767 seconds. This is because the TEA sleep function uses an int to count in ten-thousandths of a second, and 32,767 is the limit of an int variable. If you need to sleep for more than 3.2767 seconds, you will need to add multiple Sleep blocks.

The Workspace

The biggest piece of the PRP interface is taken up by the workspace, which is used to build up your program block by block. As you saw earlier, you build programs by selecting components from the palette and adding them to the workspace.

Adding a Component

We showed you how to add components to a program when we built the sample program earlier in the chapter. In this section, let's see some of the finer points of this action.

As we saw earlier, adding a component to a program can be as simple as clicking a component in the palette and then clicking on a component in the workspace. When you do this, the new component is added to the program in a position that follows the component you clicked.

Some components, however, can contain other components. These components are the If, Else, While, For, Main, and Subroutine components. When these blocks are added to the program, the PRP automatically adds an EndIf, EndElse, EndWhile, EndFor, EndMain, or EndSubroutine block.

When adding a program block, if you click the block that begins one of those components listed earlier, the new block is added between the component and its corresponding end block. Thus it becomes part of the code section contained by the component. For example, program blocks added between If and EndIf will be executed only when the condition of the If block is true. When the If section also contains an Else block, program components between Else and EndElse will be executed when the condition of the If block is false.

NOTE: The PRP will not stop you from adding program blocks between the EndElse and EndIf blocks. However, if you do so, your program may not work correctly.

If you click the workspace—not an existing component—while adding a new component, the PRP will usually ignore the mouse click, because it doesn't want to guess where you want the component to be located.

Deleting and Pasting Components

To delete a component from a program, simply click the component in the workspace, and choose Edit | Cut or Edit | Delete, or click the corresponding

Cut or Delete button from the toolbar. Cut removes a block from the program but saves a copy so that you can paste it into the program. Delete permanently removes the block from the program.

NOTE: Be careful when cutting or deleting If, Else, For, and While blocks or their corresponding End block. You must delete both the block and its corresponding End block. If you delete one and not the other, your program cannot be translated.

When you cut a program block, the block is placed into a paste buffer. To paste the block to a new location, click an existing component in the workspace to select the block and then choose Edit | Paste or click the corresponding Paste button. The new block will be pasted into the program following the selected block.

You can also copy program blocks and paste copies of them to new locations in the program. A block is copied by selecting a block and then choosing Edit | Copy or by clicking the Copy toolbar button.

Creating a Subroutine, Function, or Method

When you're creating programs with the PRP, you may sometimes need to perform the same sequence of actions, or perform the same calculation, several times within the program. Rather than adding the same components over and over again to the program, you can create a subroutine to do this for you.

We are using the term *subroutine* as a generic term to mean any number of program structures, depending on the software language. Basic does not have formal subroutines, but we can simulate them using language features. TEA has subroutines and functions; functions return a value to the caller and subroutines do not. Java uses the term *method* to mean some code that one can call; methods may or may not return values.

To add a subroutine to a program, choose Edit | Add Subroutine. The PRP will add a Subroutine and EndSubroutine block to the workspace. If you decide later that you want to delete the subroutine, select the Subroutine block and choose Edit | Delete Subroutine. In Figure 6-7, you can see a version of the program we created at the start of this chapter, but now it uses subroutines.

Figure 6-7
In this version of the To-Fro program, the ServoStart and ServoStop blocks have been moved into Subroutines. Now we can make the robot go back and forth additional times by adding Call blocks to call the subrotuines, rather than repeating the ServoStart blocks every time we want to add a movement.

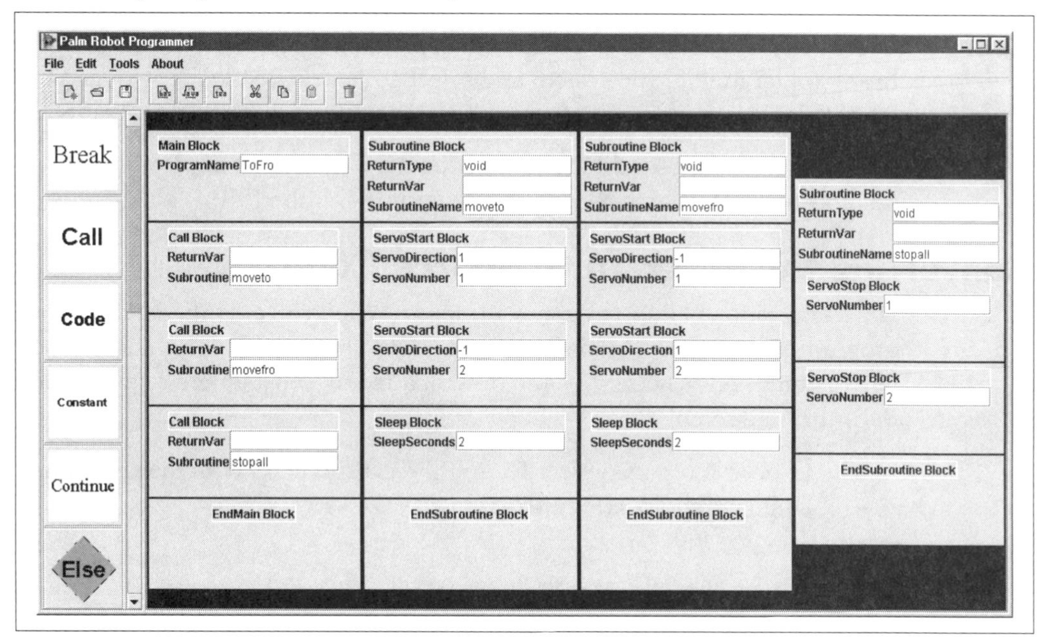

Saving and Loading Programs

After creating your program, you will want to save it on your computer for later use.

Saving

To save any program you've created, choose File | Save, or click the Save button on the toolbar.

You'll see a standard file Save dialog box. Use this dialog box to navigate to any folder on your PC. Enter a name for your program and click the Save button. The PRP will add the appropriate file extension for you. (The standard extension for PRP programs, in case you're curious or think it might come in handy on an episode of *Jeopardy*, is **.prp**.)

Loading

Loading a saved program is also straightforward. To load any program you've created, choose File | Open, or click the Open button on the toolbar.

This displays a standard file Open dialog. Using the dialog, you can navigate to any folder in your file system and select a file to be loaded. The dialog defaults to looking for files with the standard .prp extension for PRP programs.

NOTE: Do not try to edit the saved program in any other application. If you do, you will not be able to load the program with the PRP.

Converting the Program to Source Code

Now's the moment you've been waiting for! When you are satisfied that your program is complete, you need to turn it into source code. This source code will be compiled as necessary and downloaded to a host where it can be used to control the robot.

1. Choose Tools | Convert To. You'll see a submenu with the available language choices (Figure 6-8).

Figure 6-8
The first step in conversion is selecting the language for the source code.

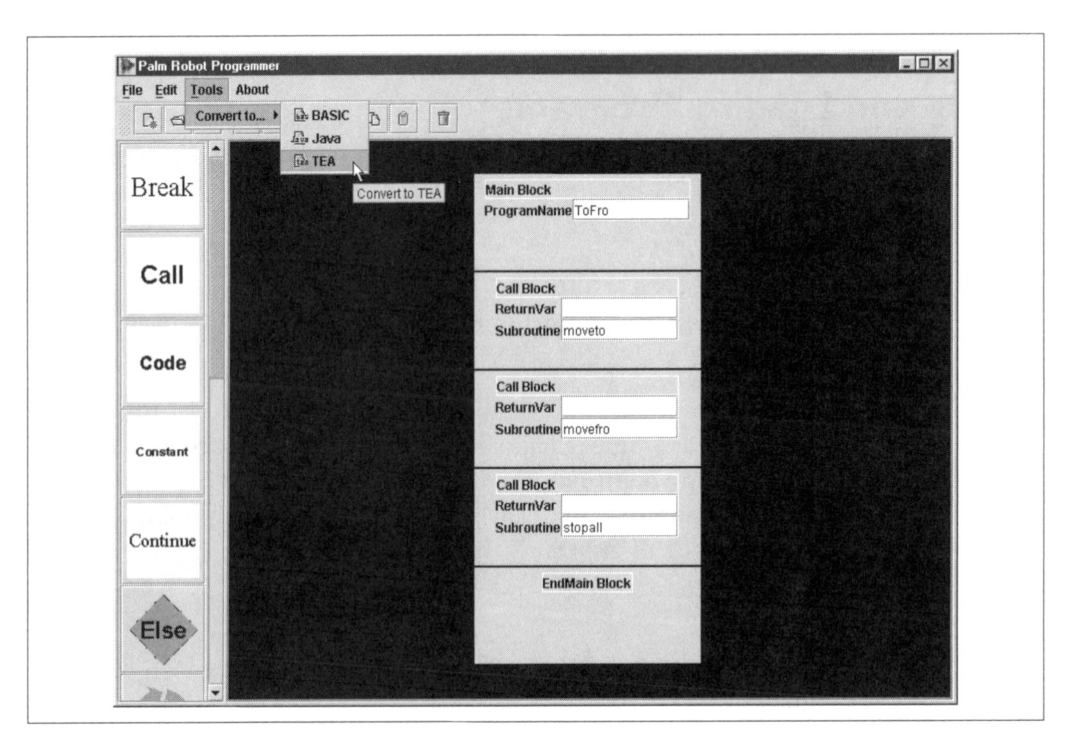

The version of the PRP provided with this book allows you to translate code into BASIC, TEA, or Java. You can use BASIC code with the Pontech SV203 controller. You will use TEA and Java with the BrainStem controller. Want other options? You'll have to be really, really nice to Kevin, because he slaved over this program for months.

2. After choosing the language for the source code, the PRP will translate the program into source code. This happens relatively quickly, and when it's complete, the program will display a dialog telling you where it saved the file.

3. You will see a dialog box if the conversion fails.

NOTE: If the conversion fails, it is probably because there is an error in your program. You will need to check your program for correctness. Check that all End blocks (If, For, and so on) have a matching beginning block. In other words, ensure each If has a matching EndIf. Do the same for all the blocks that have a matching End block.

4. After the source file is created, you may need to transfer it to a PDA or to the controller, so that it can be executed.

Loading Source Code to the Host

After you have successfully converted your program to source code, you are ready to transfer it to a host for use with the robot. The host you use depends on how you constructed your robot and what interface cables you used.

The host for the PPRK using the Pontech SV203 controller will be a Palm OS device such as a Palm III, Palm V, or Handspring Visor. The host for the PPRK using the BrainStem controller can be one of the Palm OS devices, but it can also be a PC connected directly to the robot.

Basic

The BASIC code created by the PRP is tailored for the HotPaw Basic interpreter. This interpreter was introduced in Chapter 5. HotPaw Basic is an interpreter for BASIC source code for your Palm device. The interpreter runs as an application on the PDA and can be used to execute BASIC programs stored as memos in the memo pad.

You can download and install HotPaw Basic from the CD-ROM in the directory \Other Applications\HotPaw, or from the web site www.hotpaw.com/hotpaw. The HotPaw Basic files come in a zipped archive. After unzipping the HotPaw Basic archive, you will need to hot sync the ybasic.prc and mathlib.prc files to your PDA.

Since HotPaw Basic executes programs stored as memos, you will need to download the translated program file to your PDA. The best way to do this is to open the Palm Desktop, create a memo, and copy the translated program from your computer to the memo, as shown in Figure 6-9.

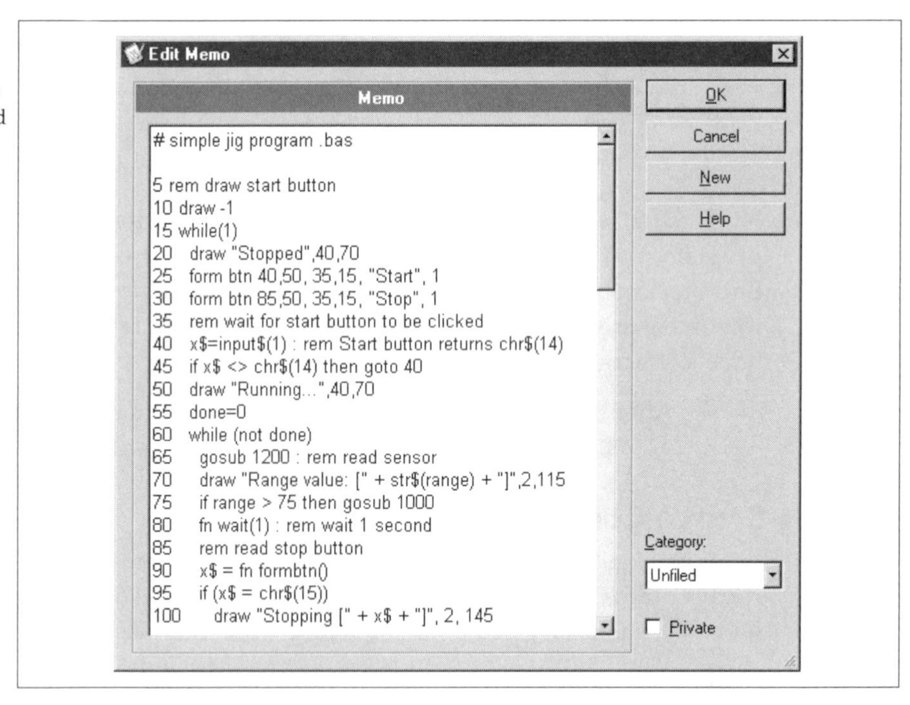

Figure 6-9 BASIC programs are easily entered into a memo in the Palm Desktop and then transferred to the PDA when you perform a hot sync operation.

The next time you perform a hot sync operation, the memo will be transferred to the PDA and ready for you to run.

1. To run your program, launch the HotPaw Basic interpreter from the Palm's Application Launcher.

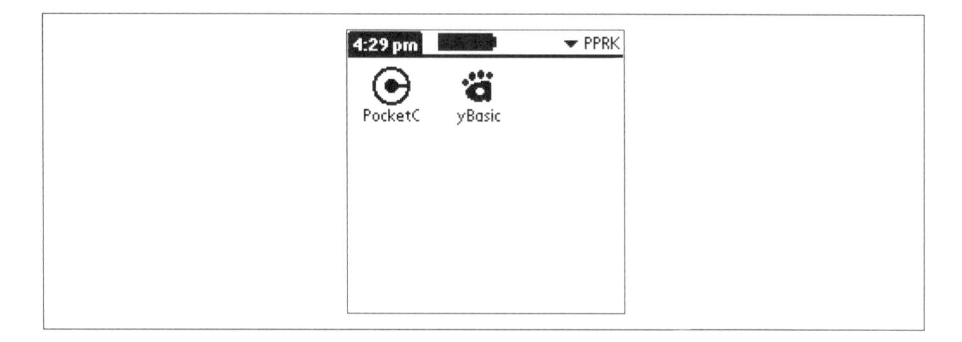

2. You will be presented with the list of BASIC programs that HotPaw found in the memo pad. Select a program and tap the Run button to run a program.

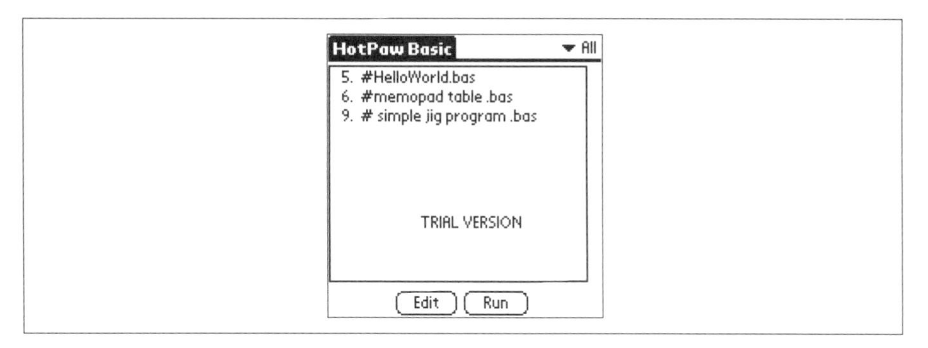

TEA

TEA, Tiny Embedded Applications, is the best language to use for the BrainStem controller. TEA code must be compiled before it can be downloaded to the BrainStem controller, and this is done with the Console program.

If you have already installed the Acroname software to your computer, you will find the Console program in the \aBinary directory. Ensure that the TEA source file you created is in the \aUser directory. You can do this by saving the source file directly to that directory when you translate the program to TEA, or you can copy it from wherever you did save it to the \aUser directory.

The TEA program is compiled using the steep command, and then it's loaded to the BrainStem using the load command. Finally, you would use the launch command to launch the program. These three steps are illustrated in Figure 6-10.

Figure 6-10 The three steps needed to launch a TEA program are to steep the source, load the code, and launch the program.

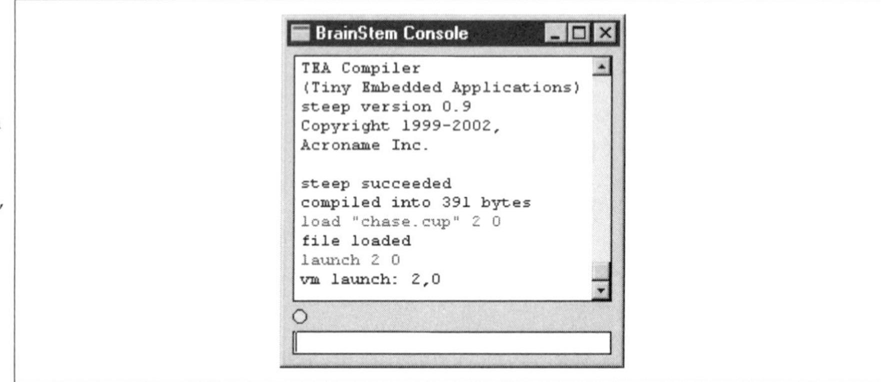

The syntax of each command follows:

- **steep "program_name.tea"** steep followed by the quoted program name with .tea extension
- **load "program_name.cup" module_id slot** load followed by the quoted program name with .cup extension, then the module ID (the default BrainStem ID is 2), and finally followed by the slot number for the program (0 to 9)
- **launch module_id slot** launch followed by the module_id (2) followed by the slot number where the program is loaded (0 to 9)

You can find additional information about TEA programs, the Console, and Console commands in Chapter 8.

Java

While using BASIC and TEA with the robot are relatively straightforward, using Java is a bit more involved. For this reason, we will not show you how to use Java with your robot here. Instead, please check Chapter 8, where we've provided detailed instructions on how to get a Java compiler, how to get a Java runtime for your PDA, and how to compile and use Java programs with your PDA and robot.

Advanced Topic: Adding New Languages

The PRP converts programs into three languages: BASIC, TEA, or Java. What do you do if you want to convert into another language? We've developed the PRP so that if you are a Java programmer, you can easily add a new language converter to the PRP.

We should emphasize that this section is intended for those of you who are already programmers and know something about Java, Extensible Markup Language (XML), and Extensible Stylesheet Language Transformations (XSLT).

How the PRP Converts a Program to Source Language

If you built the example program earlier in this chapter and saved it to your system, you can navigate to that file now and have a look at it. (If you didn't create the sample program earlier, go back and do that now.) When you convert a program to a source file, two files are created. The first file contains the translated source code for the program. The other file contains an XML representation of the program.

If you have translated a program, navigate to the directory that contains the source file. In that directory, you should find a file with the same name but with a .xml extension. Open the XML file in a text editor. Here is the XML file for the program we created at the beginning of this chapter:

```
<?xml version="1.0"?>
<PalmRobotProgram>
<Main programName="ToFro">
<ServoStart servoNumber="1" servoDirection="1"/>
<ServoStart servoNumber="2" servoDirection="-1"/>
<Sleep sleepSeconds="2"/>
<ServoStart servoNumber="1" servoDirection="-1"/>
<ServoStart servoNumber="2" servoDirection="1"/>
<Sleep sleepSeconds="2"/>
<ServoStop servoNumber="1"/>
<ServoStop servoNumber="2"/>
</Main>
</PalmRobotProgram>
```

When the PRP translates a program, it creates an XML representation of the program. All of the program constructs and icons that you create in the PRP workspace are stored in this XML file. Next, the PRP uses XSLT to translate the XML into source code. In simple terms, XSLT provides a standard way to use a style sheet to transform XML into some other representation.

How to Write Your Own Converter

To create your own converter, you will need to be able to create a relatively simple Java class and a more complicated XSLT file. After installing these files to the PRP installation, they will be recognized the next time you run the PRP, and they can be used to convert a PRP program to source code.

The Java Class

To add a new converter to the PRP, we write a simple Java class that is recognized by the PRP. Using the Java class, the PRP knows what new language selection to add to the Converter menu and which style sheet to use when translating the program to source code.

Only a few requirements for the class are necessary. The converter class must

- ❑ Extend the com.prp.AbstractTranslator class
- ❑ Provide a getName( ) method that returns the name of the language for which the translator is used
- ❑ Provide a getStyleSheet( ) method that returns the name of the style sheet to use for the translation
- ❑ Provide a translate(Program p) method
- ❑ Provide a getIconName( ) method that returns the name of the icon for the toolbar (you will also need to create an icon)

Here is a sample class for the mythical language "Cream":

```
public class CreamTranslator extends AbstractTranslator {
  public String getName() { return "Cream"; }
  public String getStyleSheet() { return "cream.xslt"; }
  public void translate(Program p) {
    String msg = "Additional message";
    doTransform(p, ".crm", msg);
  }
  public String getIconName() {
    return "Cream.gif";
  }
}
```

The getName( ) method is used by the PRP to create a menu entry for the translator. The getStyleSheet( ) method is used during the translation to identify the style sheet to use for the translation. The getIconName( ) method is required, but if you do not have an icon, the method can just return an empty string. If you do create an icon, it should be 16x16 pixels so that it is the same size as the existing icons. The PRP will provide a default icon if you do not provide one.

The translate(Program) method is a little more involved. To perform the translation, your code will call the doTransform(Program, String, String) method. This method is implemented in the AbstractTranslator class. The first string parameter is the extension to add to the translated source code file (including the leading dot). After a program is translated, the PRP displays a dialog box in which the name and location of the translated source code file is displayed. If you want to include an additional message with that dialog box, you can define that message in the translate(Program) method and pass the message as the second String parameter of the doTransform(Program, String, String) method. If you do not have an additional message, pass an empty string rather than a null as the message.

The XSLT Style Sheet

The second step to providing a new translator is to create an XSLT style sheet that is used by the translator code to convert the XML file to the source code file. For every program component in the PRP palette, the style sheet needs to convert the corresponding XML tags created by the component into the appropriate code in the new language.

To complete this step, you already need to know how to create an XSLT style sheet—teaching you how to write an XSLT style-sheet is beyond the scope of this book. In simple terms, an XSLT style sheet provides a translation for every element in an XML file. For example, here are two entries of the XSLT style sheet for the TEA language (which you can find in the PRP installation directory as tea.xslt):

```
<xsl:template match="ServoStop">
aServo_Stop(<xsl:value-of select="@servoNumber"/>);
</xsl:template>

<xsl:template match="Sleep">
aCore_Sleep(<xsl:value-of select="number(@sleepSeconds) * 10000"/>);
</xsl:template>
```

The first entry is the template for the ServoStop element. Whenever the tag <ServoStop> is found in the XML file, the XSLT converter applies the template to the tag. If you look at the XML listing, you should see a ServoStop tag that looks like this:

```
<ServoStop servoNumber="1"/>
```

The template tells the translator to replace that tag with the text *aServo_Stop(* followed by the value of the servoNumber attribute, and followed by a closing *);*. So what is written to the source code file is this:

```
aServoStop(1);
```

The second template is similar to the first. It takes a <Sleep> tag and converts it into a line of code that calls the aCore_Sleep( ) function, changing seconds to ten-thousandths of a second at the same time. Thus, this tag

```
<Sleep sleepSeconds="2"/>
```

becomes this line of code:

```
aCore_Sleep(20000);
```

The XSLT style sheet you create for your new language will need to handle all the program icons used by the PRP. In other words, every icon in the palette will have a corresponding template in the XSLT style sheet. In addition, you will need a template for the tags used for the Main and Subroutine blocks. Finally, as you can see in the XML listing earlier in this section, the root element of the XML is <PalmRobotProgram>, so your style sheet will need a template for that tag.

The Robot Geek Says

If you want to learn how to create and use XSLT style sheets, numerous books are available for you. We recommend these:

- *XSLT* by Doug Tidwell, from O'Reilly & Associates
- *XSLT Quickly* by Ron DuCharme, from Manning Publications Company

Just a couple of final steps are needed to add the new language translator to the Palm Robot Programmer. First you need to compile the Java file you created. When the file is successfully compiled into a class file, copy the class file to the \classes\com\prp\converter directory of the PRP installation. You'll know you have the correct directory when you see the existing translator classes and XSLT style sheets in the directory. If you created a toolbar icon, copy that file to the same location. Finally, copy the XSLT style sheet to the same location. The next time you run the PRP, the program will find the new translator class and add an entry to the menu for the new language.

Finally, the PRP loads the style sheet every time you translate a program. That means that if you need to debug your style sheet, you do not need to restart the PRP. Simply make the change to the style sheet, and the PRP will load the new style sheet when you next translate a program into source code.

Updates to the PRP

We all know that software changes and improves over time. (Sometimes software changes and gets worse, but that's another story.) We expect the same thing (improvements, not problems) to happen with the PRP. The CD-ROM that comes with this book has the version of the PRP as it existed when the book was published. However, you may be reading this book years later, after James Cameron has optioned it for a major motion picture and Anne Rice has written, "A real page turner—you've got to read this one for yourself."

Or, you may have lost the CD-ROM to this book.

You will always be able to find the latest version of the PRP by checking our web sites:

- Dave's web site is www.bydavejohnson.com.
- Kevin's web page is home.earthlink.net/~kmukhar.

Where to Go from Here

If you're a code slinger, we've been promising that we'd show you how to write your own code for the PPRK. You actually got a little bit of that here. But it gets better in the next chapter, where we show you how to write code for the SV203 controller, which is used in the original PPRK. We'll show you how to write code in BASIC and C, how to download that code to a PDA, and how to use that code to control the robot.

In Chapter 8, we'll show you how to write code for the BrainStem controller using the TEA language. In addition, we'll look at writing Java programs for the BrainStem. Finally, we'll show you how to download TEA or Java code to a PDA and how to use it to control the BrainStem.

Chapter 7

Essential Robot Programming Strategies

This is the chapter that all you hackers and computer geeks have been waiting for. This is where we show you how to write programs for your robot. Actually, we've divided this information into two chapters. In this chapter, we show you how to write programs for the PPRK and the Pontech SV203 controller. In the next chapter, we show you how to write programs for the PPRK and the BrainStem controller.

We will explore two languages for writing programs for the Pontech controller. We'll start by looking at writing programs in BASIC. There are a couple of BASIC interpreters that you can use. The first one we'll look at is HotPaw Basic, which runs on the Palm. BASIC programs are stored as memos on the Palm OS device. We'll also spend a little time with NS Basic. NS Basic is a forms-based programming environment for the Palm. You create Palm programs on your desktop PC by creating forms and form controls such as buttons and fields. You then add code to the controls, and the code is executed when the control is activated. Finally, we'll look at writing C programs for the Palm using PocketC. PocketC is similar to HotPaw Basic in that your C programs exist as memos on the PDA that are executed by the PocketC application. We had also wanted to include a section on programming the PPRK in Klingon, but our Klingon programmer was busy destro … that is to say, debugging some other code.

The Robot Geek Says

In this chapter, we assume that you already know how to program in at least one computer language. This chapter will not be a tutorial on how to program, but rather how to use what you already know to program for the SV203 or BrainStem controller. If you want to learn how to program, there are many other books you can try:

- *Teach Yourself C*, and *C++: A Beginner's Guide*, both by Herb Schildt, from McGraw-Hill/Osborne
- *C How to Program*, by Harvey M. Deitel and Paul J. Deitel, from Prentice Hall
- *Learn to Program tlhIngan in 24 Hours*, by Kev Martok, from Starfleet Academy Press
- *Robot Invasion*, by Dave Johnson, from McGraw-Hill/Osborne (It has some robot programming, but it's not a programming book per se. We thought that as long as we're plugging other people's books, we might as well plug one of our own.)

Programming the PPRK in BASIC

Although the developers of the PPRK used C++ to program their robot, it is much easier to get started using BASIC. With one version of BASIC that we will look at, you don't need to know any of the details of Palm OS programming, such as how to create forms and buttons. Your BASIC program does not need to be compiled, it exists as a simple text file that is easy for you to edit and store on the PDA. With many C or C++ systems, you can't edit the program on your PDA as you can with a BASIC program. It's also much easier to bail out of a badly behaving BASIC program; with C code, a bug in the code could result in a hard reset for the PDA.

So, have we convinced you to give some BASIC code a try? If so, press on. If not, we'll have our "editors" Tony and Guido come over to give you a little more persuasion.

HotPaw Basic

HotPaw Basic is the work of Ron Nicholson. It is a full-featured BASIC programming system that runs on your PDA. Programs are stored as memos in the MemoPad, and a simple tap of the stylus will run any program. Because the program is stored as a memo, it is easy to edit the program right on your PDA. The web site for HotPaw Basic is www.hotpaw.com/hotpaw.

Download HotPaw Basic from the Web or from the CD-ROM and unzip it onto your hard drive. There are two files that you need to load to the PDA: ybasic.prc and mathlib.prc. The yBasic application is what runs your BASIC programs. The file mathlib.prc is a library of double-precision math routines used by HotPaw Basic.

Find and tap the icon for yBasic in the Applications Launcher.

This will bring up a program list form that shows all the programs that HotPaw Basic found in the MemoPad. If you just installed HotPaw Basic, this list will be empty and the message "No BASIC programs were found in the MemoPad" will be displayed in the form. Below the form are two buttons. When programs are listed in the form, you can select the program and use the buttons to edit or run the selected program.

Hello, World!

Let's start with a simple example that will show you some of the features of HotPaw Basic. In this example, we'll draw a form on the screen and use some buttons to accept input from the form. For now, don't worry about understanding every keyword or function in the listing. We just want to get a quick example working. This program is short enough that you could enter it directly into a memo on your PDA. However, most of the time it will be easier to type the memo into the Palm Desktop and then hot sync it to the PDA.

What do we want the program to do? First, it will display the phrase "Hello, World!" on the screen. Next, it will draw a button on the screen.

Every time the user taps the button, the program will display a message box that asks the user to enter a name.

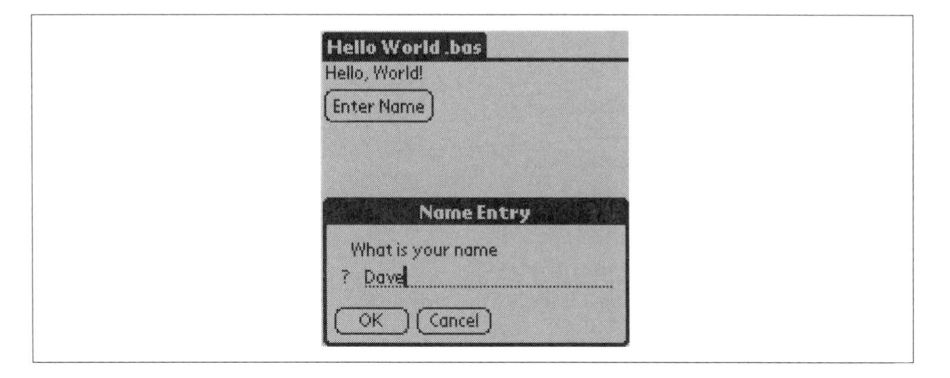

If the user enters a non-empty string, the program will change the displayed phrase to "Hello, *name*."

To find programs, HotPaw looks for memos in the MemoPad with a first line that begins with the number character (#), followed by some text that ends with .bas. So here's the first line of our program:

```
#HelloWorld.bas
```

Note that even though there are no space characters in the line above, HotPaw Basic doesn't care how many letters, numbers, or other characters are in the first line, as long as it begins with # and ends with .bas.

Next, we'll create some variables. Since these are string variables, the variable name includes a trailing `$`. Note that you can put multiple code statements on a single line if they are separated by a colon.

```
a$ = "Hello, " : b$ = "World!"
```

Now we'll create a loop so that the program runs until we tell it to quit. Inside the loop, we'll draw the phrase to the screen:

```
while (1) : rem loop forever
draw -1 : rem clear the screen
draw a$ + b$,2,15
```

There are two commands for printing to the screen, print and draw. With the draw command, you pass the string, followed by the screen position in x and y coordinates where the upper-left corner of the screen is 0,0.

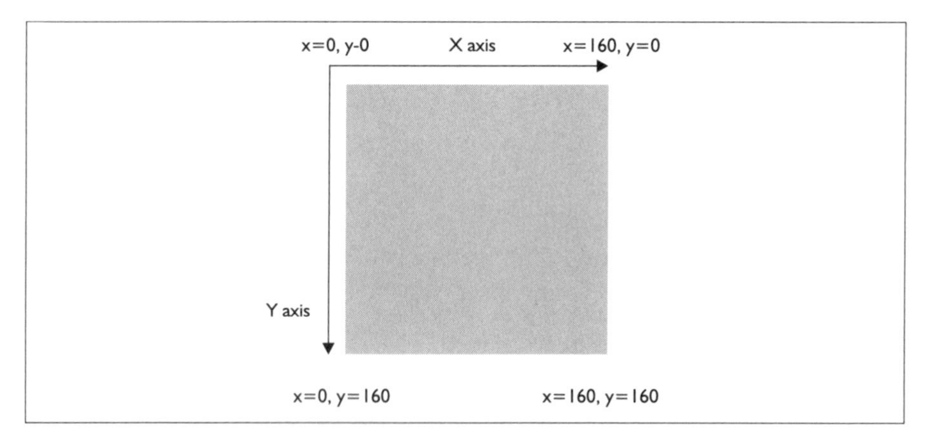

The line of code above draws the string "Hello, World!" at screen position x=2 and y=15, which is just below the application title. Now, let's create a form button that will pop up an input box. The command for creating a form button is form btn x, y, width, height, button_label, 1.

```
form btn 2,30, 55,15, "Enter Name", 1
```

This draws a button that is 55 pixels wide and 15 pixels high at screen position x=2, y=30. When the user taps the button, we'll pop up an input box that asks the user for his name. When the program jumps back to the start of the loop, the program prints "Hello," with the name that was entered.

```
x$ = input$(1) : rem wait for button press
i$ = inputbox("What is your name", "Name Entry", 2)
```

The inputbox( ) function can take three arguments. The first is a query string displayed in the box, and the second is a form title. The final argument, 2, tells the function to display both OK and Cancel buttons. Finally, the program checks that the user has actually entered something. If the entered string is not empty, it is saved in variable b$. When the program loops back to the start, it will draw the phrase using the new value of b$.

```
if i$ <> "" then b$ = i$ : rem if user entered something, store it in b$
wend : rem end of loop, go back to start
```

Here is the complete program listing:

```
#HelloWorld.bas
a$ = "Hello, " : b$ = "World!"
while (1) : rem loop forever
draw -1 : rem clear the screen
draw a$ + b$,2,15
form btn 2,30, 55,15, "Enter Name", 1
x$ = input$(1) : rem wait for button press
i$ = inputbox("What is your name", "Name Entry", 2)
if i$ <> "" then b$ = i$ : rem if user entered something, store it in b$
wend : rem end of loop, go back to start
end
```

There are a few more things we want to point out about the code above. First, while BASIC programs are usually line-numbered, the code above is not. HotPaw Basic will automatically apply line numbers internally. Line numbers can be useful when you are debugging the code and when you want to create subroutines. We'll do that in a later program. Also, comments in the code start with the keyword rem. Anything that follows rem up to the end of the line is a comment and is not executed. Finally, the program ends with the keyword end. The end keyword is not really necessary in this program because it runs in an endless loop. However, it is useful to get in the habit of putting an **end** at the end of your main program. Later in this chapter, we will write a HotPaw Basic program that uses subroutines that follow the main program.

Without the end, a program that does not run in an endless loop could easily run past the end and start executing the subroutines, which would be a program bug.

When you are finished entering the program onto your PDA (or loading it from the Palm Desktop), find and tap the yBasic icon in the Application Launcher. This time the list of programs should show, #HelloWorld.bas. Select the program and then tap the Run button.

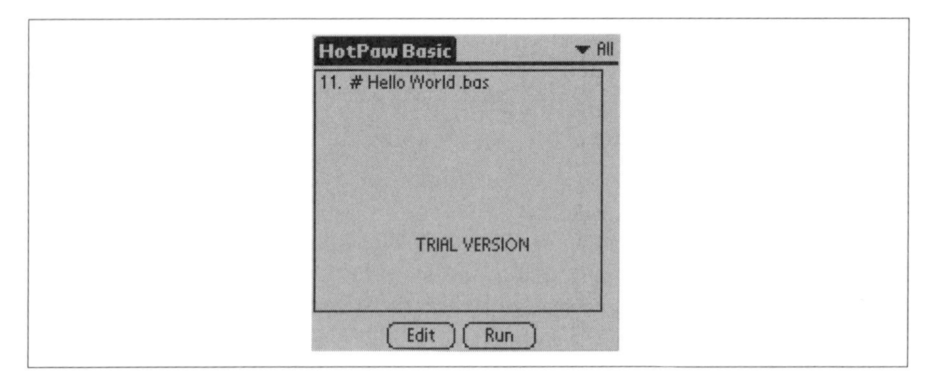

You should see the text "Hello, World!" printed to the screen with a button that says Enter Name below the text. Notice also that HotPaw automatically puts a Quit button at the bottom of the screen. This is what allows us to get out of the infinite loop that is in the code above. When you are finished, tap the Quit button. You will now be back at the program listing screen. If you select the #HelloWorld.bas program, you can tap the Edit button to edit the memo.

Some Useful HotPaw Commands and Functions

HotPaw Basic makes programming a Palm easy because you don't have to worry about the details of Palm OS programming. One of those details is all the work in creating forms and form controls such as buttons, message boxes, labels, and so on. Using simple commands or functions, HotPaw handles all the details for you. Table 7-1 lists some of the more common and useful form controls you can use with HotPaw Basic. To find complete details about how to use all the commands and functions of HotPaw Basic, you'll want to look at quickref.txt and yb_tutorial.txt, both of which come with the software. The file quickref.txt lists all the commands, functions, and keywords you need to know to write BASIC programs for the Palm OS. The file yb_tutorial.txt contains many sample programs showing how to use HotPaw Basic.

Command	Description
msgbox(message$)	Displays a message dialog box with a message string and an OK button. The text message will word-wrap up to two lines. You can force line wrapping up to three lines by embedding chr$(10) chars in the message$ string.
msgbox(message$, title$)	Displays a message dialog box with a message string, the given title, and an OK button.
msgbox(message$, title$, n)	Displays a message box with a message string and title. If n = 2, display OK and Cancel buttons; otherwise, display an OK button.
print a,b	Display a and b in a message box, similar to the msgbox subroutine. Provided for backwards compatibility with older versions of HotPaw Basic. New programs should use msgbox.
?	Same as print command. Provided for backwards compatibility with older versions of HotPaw Basic. New programs should use msgbox.
print at x,y	Sets print-to-window-mode for all following print statements. To turn off print-to-window mode (in other words, to print in message box mode), use print at -1,0.
input$(prompt$)	Displays an input dialog box with the given prompt.
input$(prompt$, title$)	Displays an input dialog box with the given prompt and title.
input$(prompt$, title$, default$)	Displays an input dialog box with the given prompt, title, and default value.
input$(prompt$, title$, n)	Displays an input dialog box with the given prompt and title. If n = 1, only display OK button; if n = 2, display both OK and Cancel.
val(string$)	Converts a string to a number. Thus, you can read a number entered by a user with n = val(input$(...)).
inputbox(...)	Same as input$(...).
input$(1)	Waits for one graffiti char or button press.
form(9,n,title$)	Displays n line 2 column form. Valid values for n are 2 to 9. HotPaw has a built-in string array s$. The first column of the form displays the strings at the even numbered indexes of s$ as labels. That is, s$(0), s$(2), up to s$(16) can be set with string values, and these values are the labels displayed in the first column of the form. The second column of the form accepts values entered by the user. Default values for each line of the form are held in the odd numbered indexes of s$. That is, s$(1), s$(3), up to s$(17) hold the default values displayed in column 2 of the form. The values entered by the user are also stored in s$(1) through s$(17). The form also has four buttons: OK, Cancel, A, and B. The return value of the form when the user taps a button is 1 for OK, 2 for the A button, 3 for the B button.
form(0)	Returns last dialog box button status and clears the button status to 0.

Table 7-1
Some of the More Common or Useful HotPaw Functions

Command	Description
form btn x,y, width, height, label$, 1	Creates a form button at the given x,y coordinate and with the given width and height. The label of the button is given by the string parameter.
fn formbtn( )	Returns the key code of the last form button that was tapped.
draw -1	Clears the display.
draw t$, x,y	Draws text t$ at given screen position.
fn wait(n)	Delays for n seconds.

Table 7-1
Some of the More Common or Useful HotPaw Functions *(continued)*

Creating and drawing forms is fun and leads to strong minds and bodies, but to work with the PPRK, we need to be able to communicate over the serial port. Some of the commands for sending and receiving data over the serial port are shown in Table 7-2.

Command	Description
open "com1:",9600 as #5	Opens the serial port. Note that you must always use the command as shown. COM1 is the only valid serial port, 9600 baud is the speed that the Pontech SV203 controller is using, and only descriptor #5 is valid for serial port communication.
print #5, a$	Prints the string a$ to the serial port.
get$(#5, n)	Returns a string up to n bytes long.
input #5, a$	Waits for one line of text.
close #5	Closes serial port.
fn serial(5)	Returns the number of bytes to be read from descriptor #5.

Table 7-2
HotPaw Basic Commands for Using the Serial Port of the Palm

PPRK Programming with HotPaw Basic

Now that we've seen a simple application using HotPaw Basic and some of the HotPaw Basic commands and functions, let's actually write some code for the PPRK. We'll first discuss what the program is intended to do, and then show the code.

The Simple Jig Program

This program will be simple in concept, but as you will see, it does involve a bit more code than we saw earlier. The program will run in a loop. Within the loop, it will periodically ask the SV203 if sensor 1 has detected an object. If it has, the program will command the robot to move away from the object, spin around in one direction, spin around in the opposite direction, and then move back to where it was. If you read Chapter 5, you may notice that this program does the same thing as the simple.tea program we looked at there. By showing you the same program in a different language, you can compare this program to the same program written in TEA. Then you can write a 500-word essay comparing and contrasting the two programs. You must have five paragraphs at a minimum, consisting of introduction, three body paragraphs, and a conclusion, with topic sentences for each paragraph. Essays are due next Monday. Send them to Dave for grading.

After we show the complete program listing, we will look at its various parts in a little more detail. This program is a bit more involved than the earlier Hello World program, so you may want to enter it as a memo in the Palm Desktop and then perform a hot sync operation to load it to your PDA. So, without further ado, here is the listing for simple jig program .bas. If you don't want to type this into your Palm OS device, we've included the source code in the \Other Applications\SimpleJig directory on the CD-ROM.

```
# simple jig program .bas

10 open "com1:", 9600 as #5
15 rem draw start button
20 draw -1
25 draw "Stopped",40,70
30 form btn 40,50, 35,15, "Start", 1
35 rem wait for start button to be clicked
40 x$=input$(1) : rem Start button returns chr$(14)
45 if x$ <> chr$(14) then goto 40
50 draw "Running...",40,70
55 range = -1
60 while (1)
65   gosub 1200 : rem read sensor
70   draw "Range value: [" + str$(range) + "]        ",2,115
75   if range > 55 then gosub 1000
80   fn wait(1) : rem wait 1 second
85 wend
90 close #5
95 end

1000 : rem jig function
```

```
rem   1 = full reverse
rem 128 = stop (centered)
rem 255 = full forward

sv1$="128" : sv2$="1" : sv3$="255"
gosub 1100
fn wait(2)

sv1$="255" : sv2$="255" : sv3$="255"
gosub 1100
fn wait(3)

sv1$="128" : sv2$="1" : sv3$="255"
gosub 1100
fn wait(2)

sv1$="1" : sv2$="1" : sv3$="1"
gosub 1100
fn wait(3)

sv1$="128" : sv2$="128" : sv3$="128"
gosub 1100

return

1100 : rem servo control
rem   1 = full reverse
rem 128 = stop (centered)
rem 255 = full forward
print #5, "SV1M" + sv1$
print #5, "SV2M" + sv2$
print #5, "SV3M" + sv3$
return

1200 : rem read sensor 1
range = -1
print #5, "BD1AD1" + chr$(13) : rem tell board sensor 1
s = fn serial(5)
if s > 0 then
  r$ = get$(#5, s) : rem read bytes
  range = val(r$)
endif
return
```

Simple Jig Explained

The program starts at line 10 by opening the serial port. The serial port will be closed if the program reaches line 90, or if the program is terminated. HotPaw automatically closes the serial port if a program is terminated. In general, you will want to open the serial port as the first action your program takes, and close the serial port only when the program ends. In line 30, the program uses the form btn command to create a form button with the label Start. In HotPaw Basic, each button is assigned a key code. The first button to be created is assigned key code 14 by default, the next gets 15, and so on. HotPaw does provide a command to change these key codes within the program, if you need to. Next, it uses the input$(1) function to wait for the user to tap the Start button.

When the user taps the Start button, the program prints the string "Running..." and begins its main loop. It begins in line 65 by calling the subroutine to read sensor 1. If the value returned is greater than 55, the program calls the subroutine to do the jig. (The board will return a digital value between 0 and 255 for the sensor reading. The closer the detected object to the sensor, the higher the returned value.) The program waits for one second, then starts the loop again.

The first subroutine, which starts at line 1000, moves the robot in its little jig. It does this by commanding the servos. With the SV203, you send a text command for each servo. The text command looks like this: SVnMn. The first n is the servo number; the second is the servo position. The position can be any number from 1 to 255. The value 128 is center or stop, 1 is full reverse, and 255 is full forward. So to command servo 1 to go forward, the command is SV1M255. The jig subroutine sets various values for three strings that correspond to the three servos, and calls the subroutine at line 1100 to actually send the command strings.

The final subroutine, at line 1200, commands the SV203 to read sensor 1, reads the sensor value if one exists, and stores it in the *range* variable. The command for reading an analog port is AD1. The program sends the command BD1 AD1 to tell SV203 board 1 to read analog port 1. (The BD command is used when one has multiple SV203s connected together, in which case each board has its own ID number. Thus, it's not really needed every time we send the AD1 command.) The fn serial(5) function returns the number of bytes waiting at the serial port. If the function returns a value greater than zero, the bytes are read as a string with the get$(#5,n) function, where n is the number of bytes to read. If the port was read, the bytes are converted into a value and the function returns.

Running Simple Jig

When you are finished entering the program onto your PDA (or loading it from the Palm Desktop), find and tap the yBasic icon in the Application Launcher. This time the list of programs should show # simple jig program .bas.

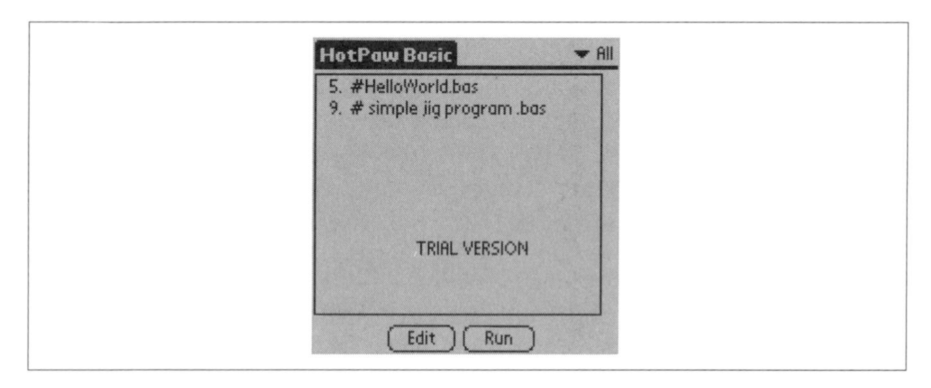

Select the program and then tap the Run button. With the PDA attached to the PPRK, turn the robot on. Then tap the Start button of the program. Place the robot on the ground and place an object (a foot works well for this) in front of sensor 1, about a foot away from the sensor. When the sensor detects the object, the robot will perform its little jig.

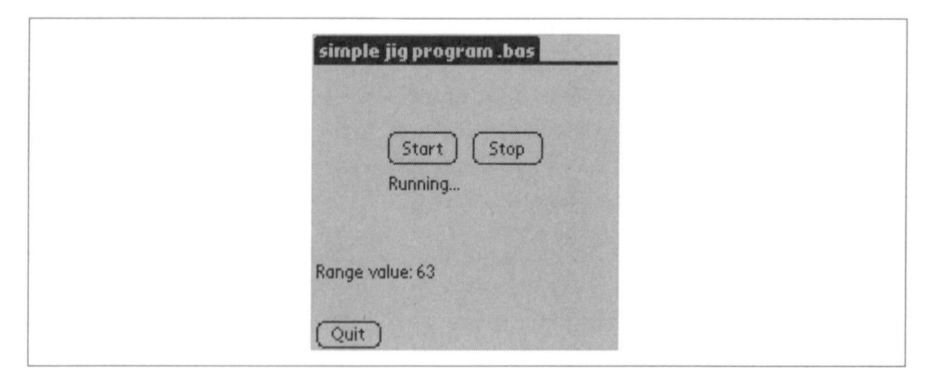

NS Basic

NS Basic, from NS Basic Corporation, is a visual integrated development environment (IDE) that runs on a Windows PC. In a visual development environment you create Palm applications by dragging controls onto forms, and then adding code to the form or control. The fact that it is an IDE means that you can write, compile, run, and debug your code without needing to switch between four different applications (editor, compiler, runtime, and debugger). Editing and compiling is done with the NS Basic application; running and debugging occurs in a Palm OS emulator (POSE). Although the two are separate applications, NS Basic is integrated with the emulator so that Palm applications can be automatically downloaded to the POSE and launched by NS Basic. The web site for NS Basic is www.nsbasic.com.

Download NS Basic from the Web or from the CD-ROM and install it onto your hard drive. The installation program will place an entry in the Start menu for NS Basic. Launch the program and you are ready to start creating a project. A project refers to the forms, resources, and code that are part of a Palm application.

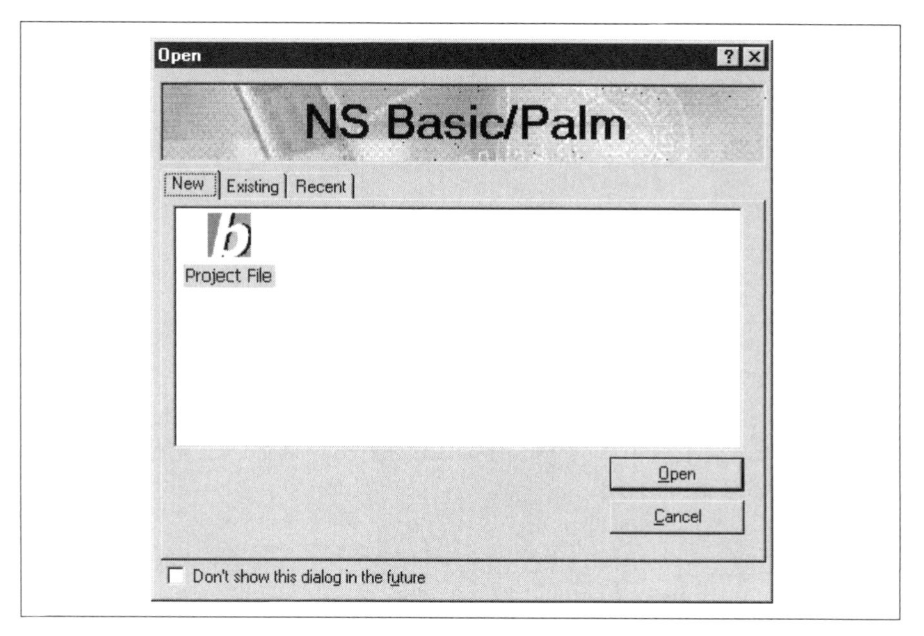

Hello World in NS Basic

Let's re-create the simple Hello World program from the previous section, using NS Basic so you can see how to add forms, controls, and code to create a simple application. Launch NS Basic and create a new project by clicking the Open button as seen in the previous illustration. The new project will open with an initial form in the center of the window. To the right is the Project Explorer pane and the Properties pane. To the left is a palette of form controls.

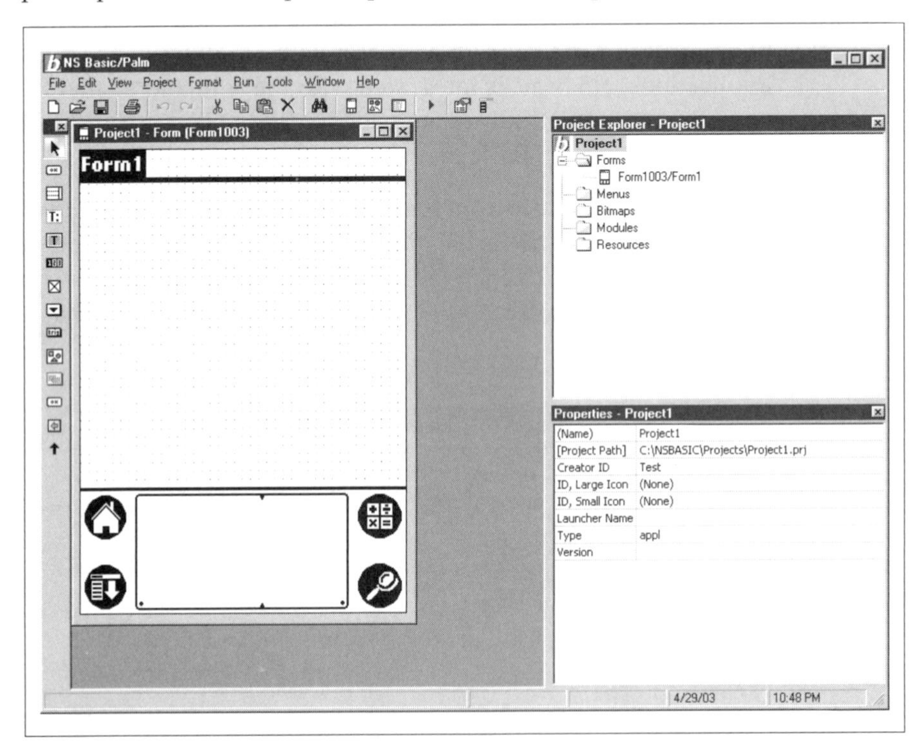

In the Project Explorer pane, click the project and the form and change their properties in the Properties pane so that the name of the project and the form is Hello World.

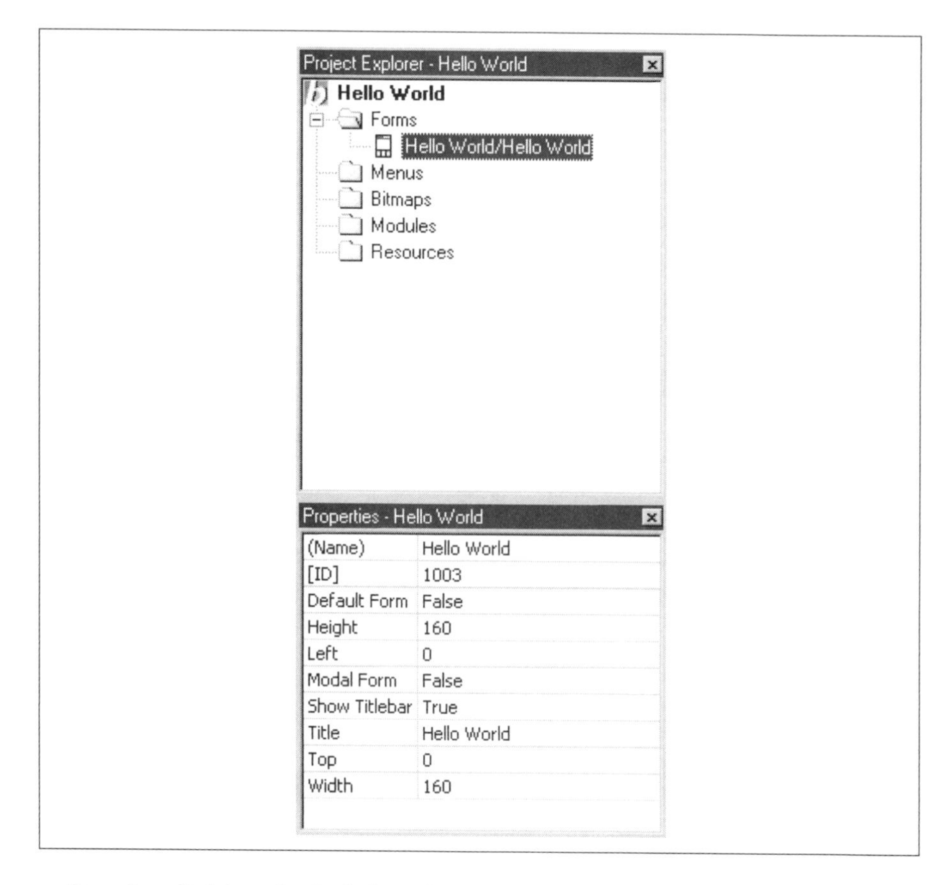

Start by clicking the Label tool in the control palette and clicking on the form in the upper-left part of the form. Click the button tool and add a button below the label. Don't worry if the names assigned by NS Basic to your controls are different from those shown in the illustration.

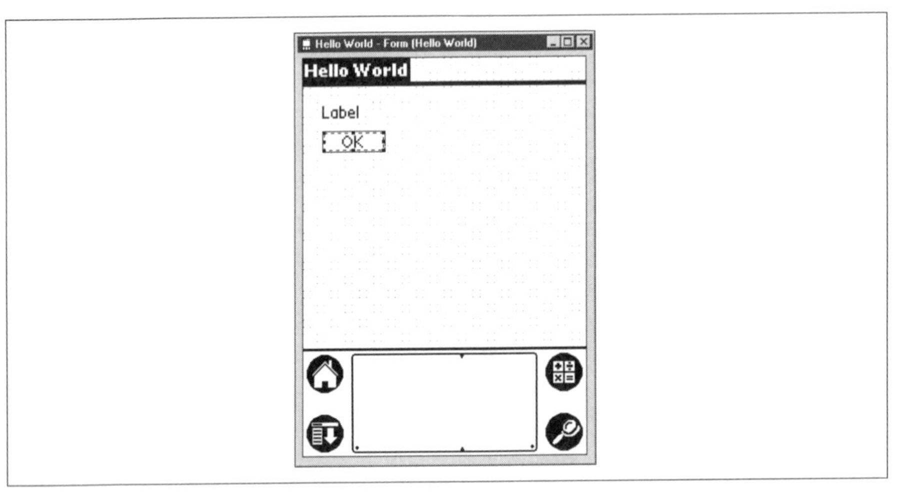

Now click the label on the form to select it. Its properties will be displayed in the Properties pane to the right. In the Properties pane, change the label to be "Hello, World!" Click the button to select it, and change its label in the Properties pane to "Enter Name."

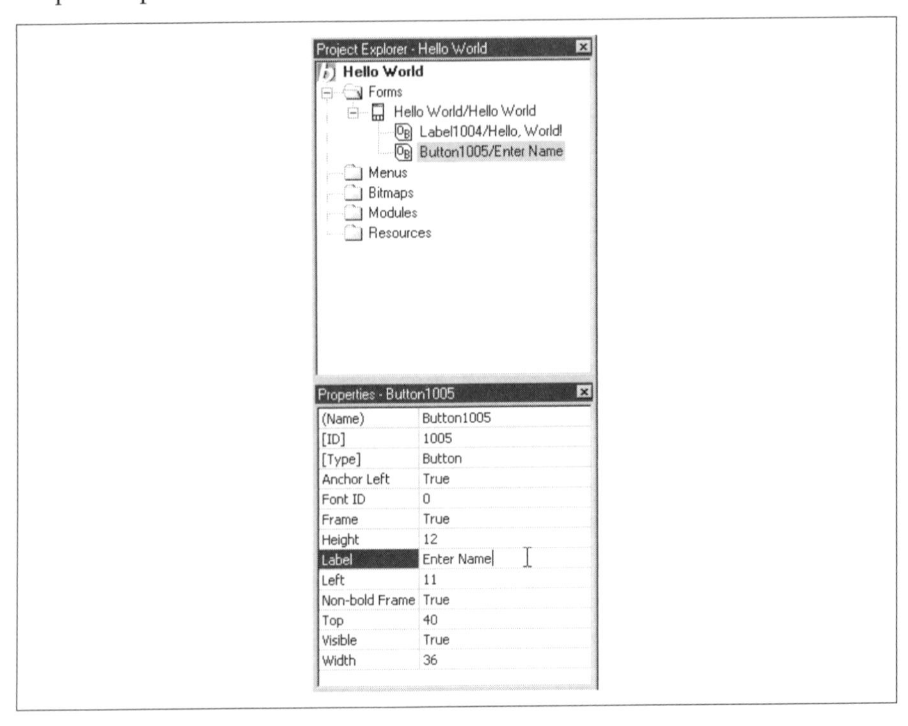

Create another label, a field, and two buttons as shown in the next illustration. Change the label text to "Enter Your Name" and the second button to "Cancel". Finally, in the Properties pane for each of these controls, set its Visible property to False.

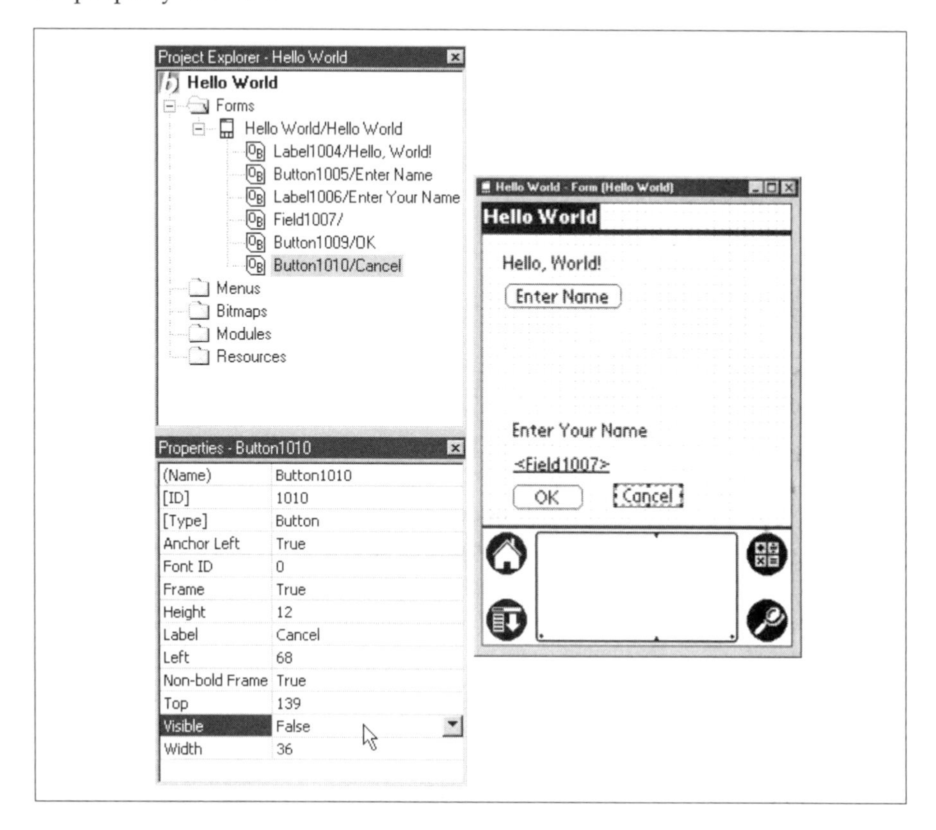

Now we're going to enter some code to do some actual work. Double-clicking a control on a form will bring up a code window. Code entered into the window will execute when the control is activated. So, if we add code to a button, the code will execute when the button is clicked. Double-click the Enter

Name button and enter this code in its code window (note that the Sub and End Sub lines are already entered into the form):

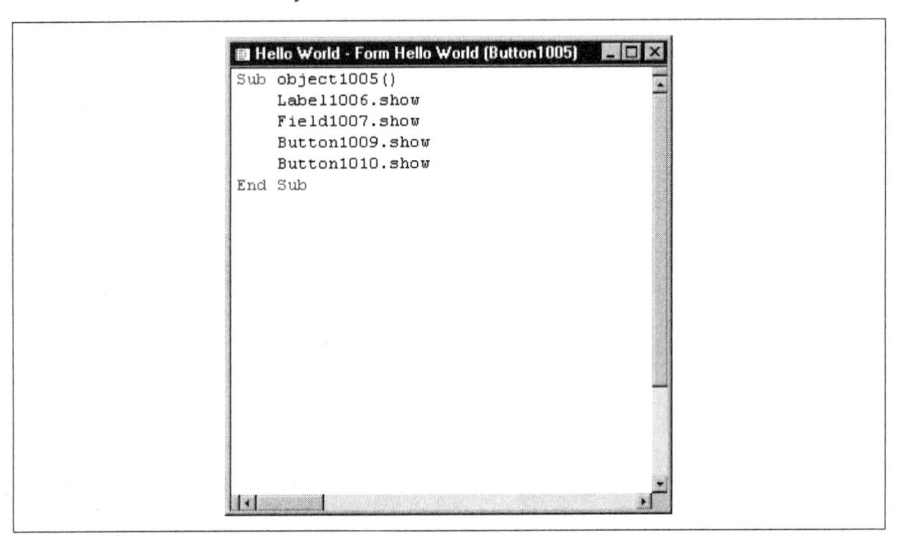

Modify the code as needed to use the correct names for the controls on the form you created. For example, if the field on your form is named, field1010, the correct code for your project would be

```
Field1010.show
```

Similarly, enter code for the OK and Cancel buttons as shown in the following illustrations.

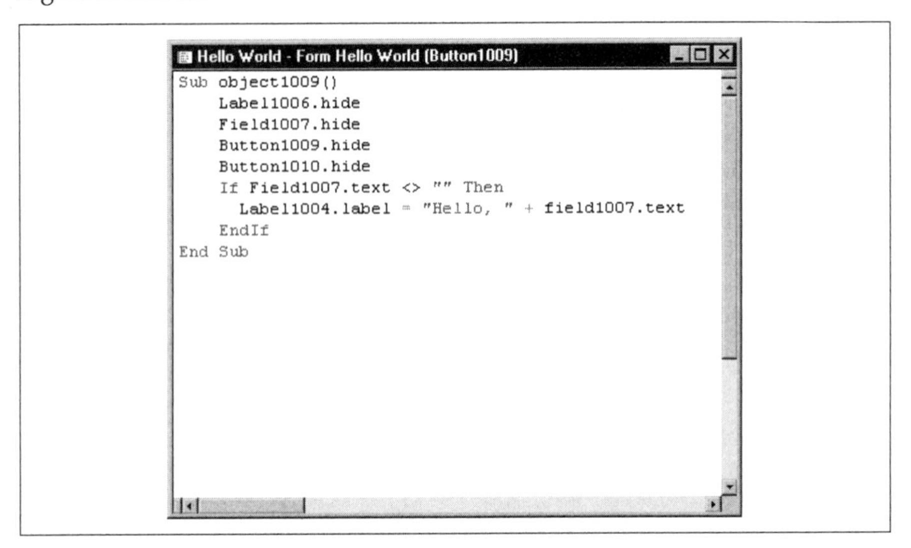

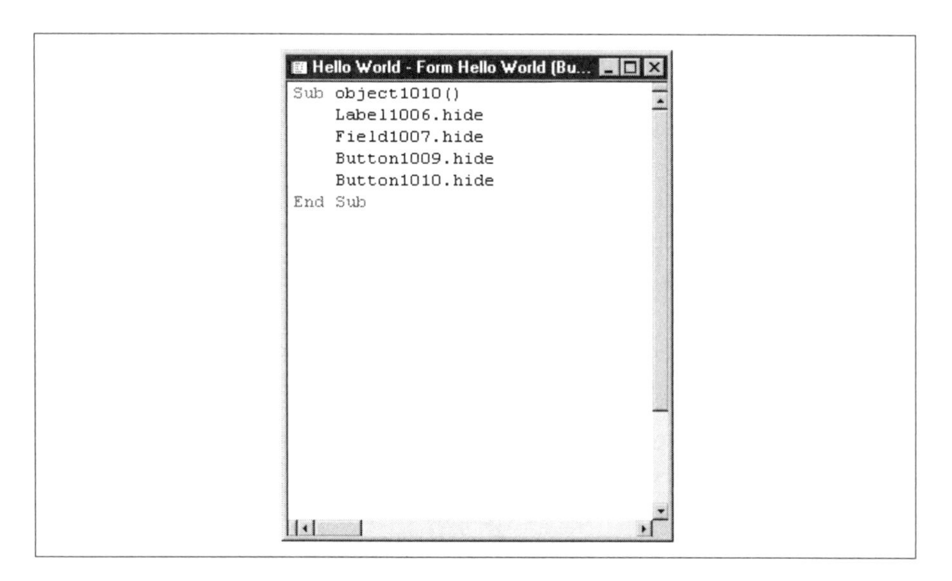

Now we need to do a little configuring of NS Basic. Select Tools | Options from the menu. Click the Compile/Download tab. Ensure that Compile Into Fat App is *not* selected. (This property seems to be reset every time you run NS Basic, so be sure to check it when you start a new project.) Also, select the option you want in the After Compiling pane. We usually choose Do Nothing so that we have control over when and how the application is launched. You may prefer to Send To POSE or hot sync To Device.

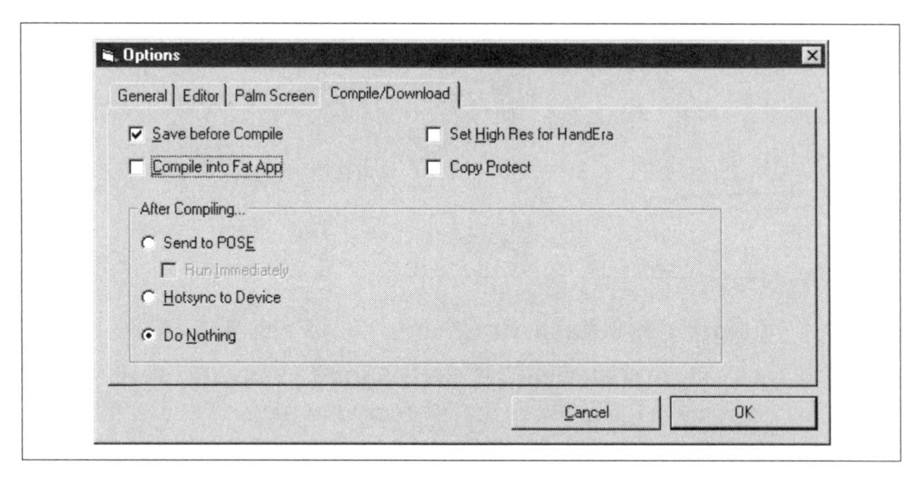

Click the OK button. Now select Run | Compile Hello World from the menu. Alternately, you can press the F5 button to perform the same action. If your project compiles successfully, you will see a dialog box that tells you the size of your program. NS Basic compiles your program into a Palm PRC file and stores it into the \Download directory of the NS Basic installation.

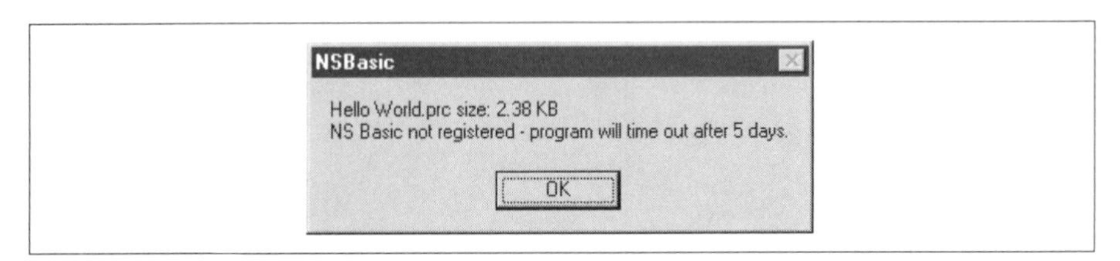

A few steps previously, we set NS Basic to create a skinny application. With a skinny application, the PRC file that is created cannot be run on its own. It needs the NS Basic runtime to be on the PDA to be able to execute. A fat application, on the other hand, is a stand-alone PRC that contains all the NS Basic runtime code as part of the PRC. To run your skinny application, you need to download or hot sync the NS Basic runtime and other libraries to the POSE or your PDA. That can be done in the Run menu.

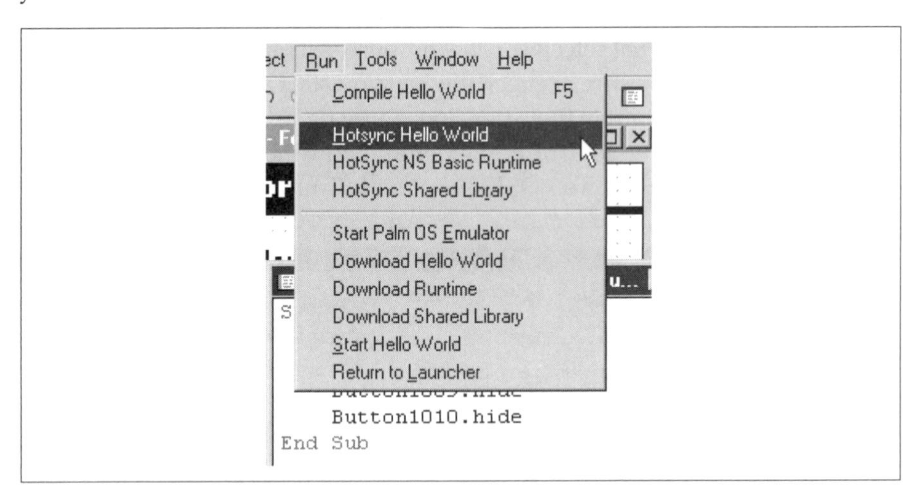

Loading NS Basic Programs to a Palm OS PDA

Select Run | Hotsync NS Basic Runtime from the menu. This will add the runtime to the hot sync list for your PDA. If you have multiple user names on your system, you will need to select the correct PDA user name for hot syncing. This only needs to be done once, regardless of how many NS Basic programs you hot sync to the PDA. Similarly, add the Hello World application to the hot sync list. The next time you hot sync the PDA, the two applications will be downloaded to the PDA. Alternately, you can use the Palm Install Tool application to install the NS Basic Runtime and the Hello World application the same way you would any other Palm application. You will find both of them in the \Download directory of the NS Basic installation.

After hot syncing the Hello World application to the PDA, find it in the Applications Launcher and tap its icon to launch it. NS Basic automatically adds a program icon to the PRC file for you, as you can see in the next illustration. (You can add your own custom icon, but we don't have room to cover that here.)

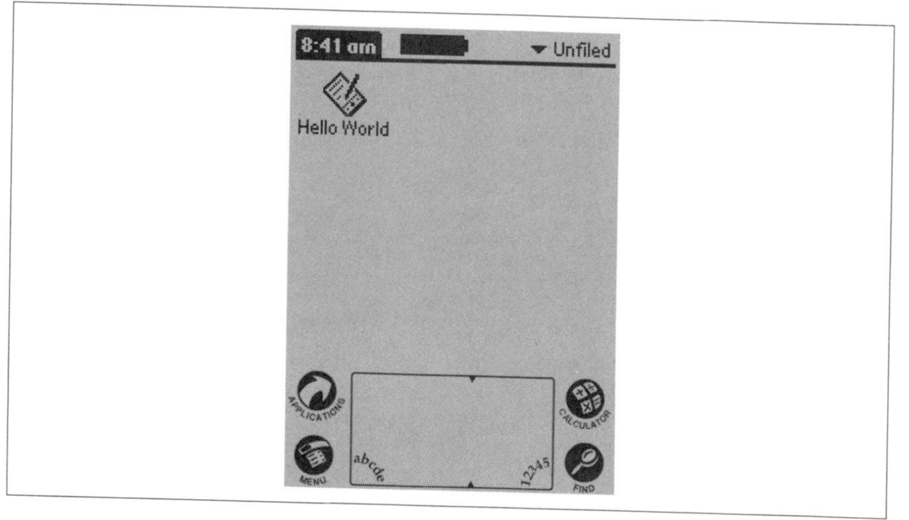

After the program launches, your PDA should look like the next illustration.

Tapping the Enter Name button will cause the other controls to become visible.

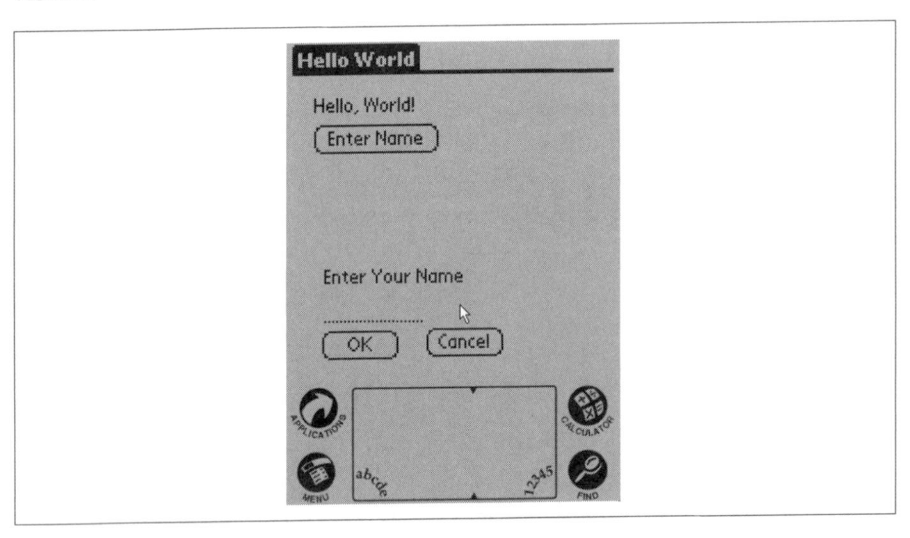

Enter a name and tap the OK button to change the Hello, World! message to be Hello, *name* and to hide the other controls again.

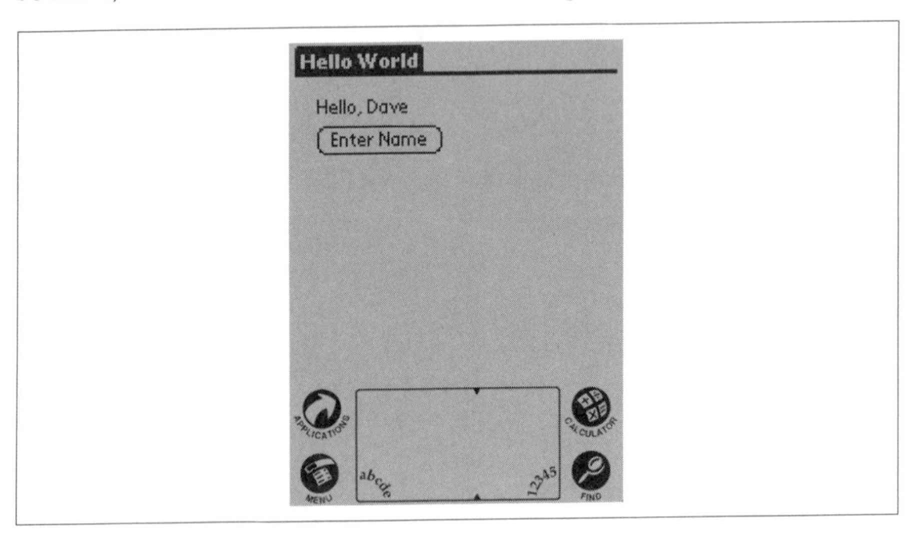

Running the POSE

Normally, you will not want to hot sync your application directly to a PDA without testing it first. That is what the POSE is for. You will need to do a little setup before you can use the POSE. Click Tools | EMULATOR from the menu to launch the POSE.

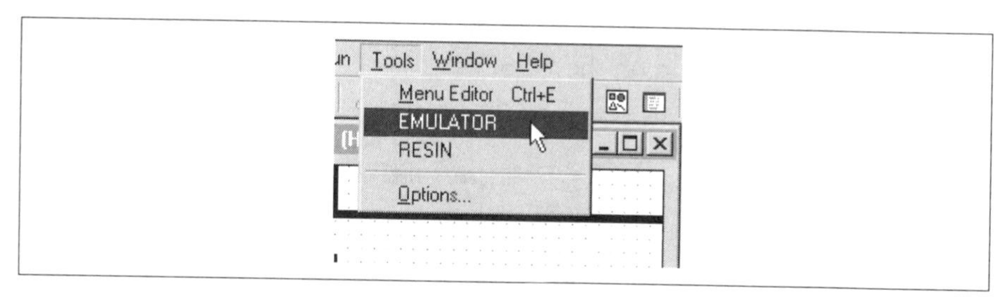

After the emulator starts up, right-click anywhere on the POSE to see the popup menu. About two-thirds of the way down is the Transfer ROM menu option. Select that option and follow the directions to download a copy of your PDA's read-only memory to your PC. After the ROM has been downloaded and saved, bring up the POSE menu again and select New from the menu. This will display the dialog box shown in the next illustration.

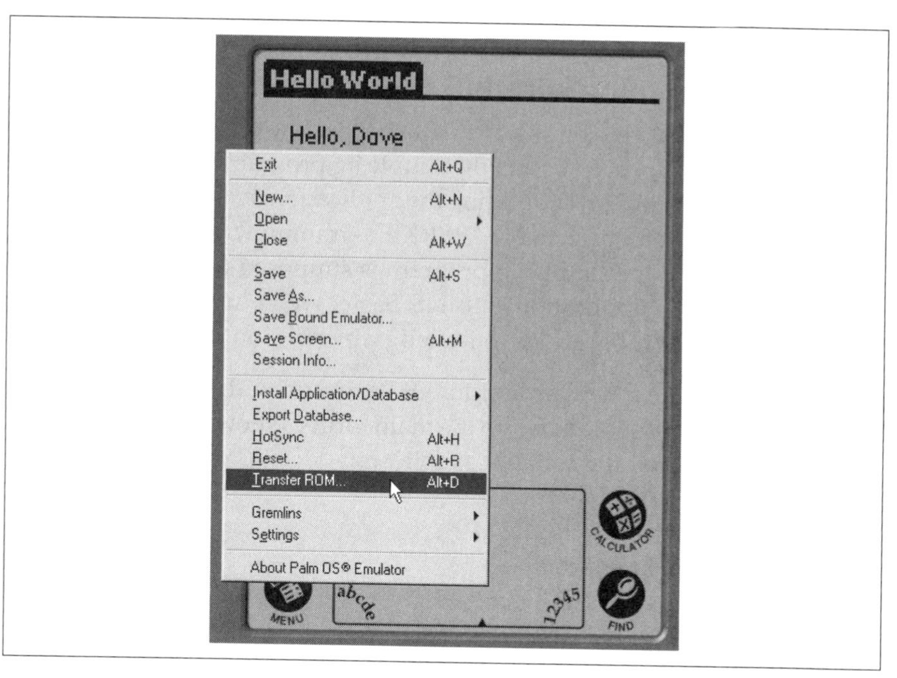

Using the drop-down selection boxes, select the ROM you just saved, select the appropriate Palm type and memory amount, and click OK to start a new POSE session.

At this point, you can use the Run menu to download the NS Basic runtime and the Hello World application to the POSE. After they are downloaded, choose Reset from the POSE menu. You should then be able to find and launch the application on the POSE.

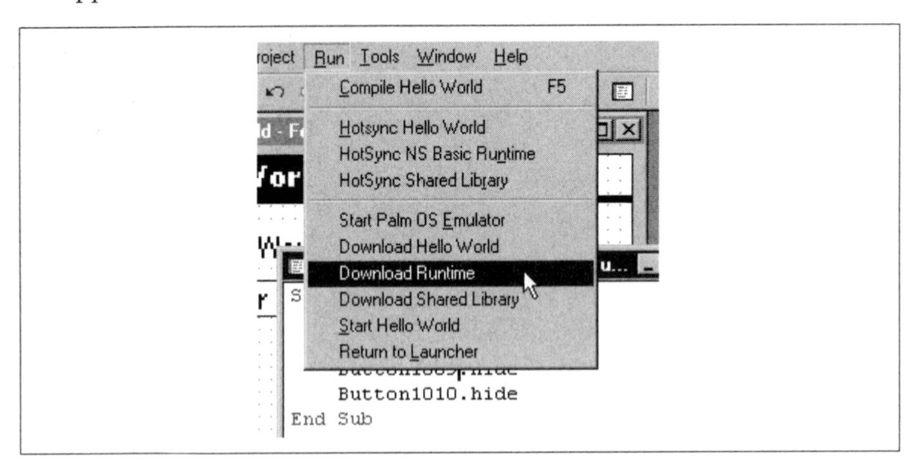

PPRK Programming with NS Basic

Now that you know how to use NS Basic to create simple forms, controls, and applications, let's redo the simple jig program in NS Basic. In this section we will quickly cover building the application using the forms and controls in NS Basic. Refer back to the "PPRK Programming with HotPaw Basic" section to see how the simple jig program is supposed to function. If you just want to load the application into NS Basic, we've included the project file in the \Other Applications\SimpleJig directory on the CD-ROM.

Create a new project in NS Basic and add two buttons named Start and Stop, and three labels to the main form as shown in the following illustration. We named the controls as follows:

- btnStart
- btnStop

- ❑ LabelStatus
- ❑ LabelRange
- ❑ LabelNumBytes

Next to the labels LabelRange and LabelNumBytes, add two blank labels.

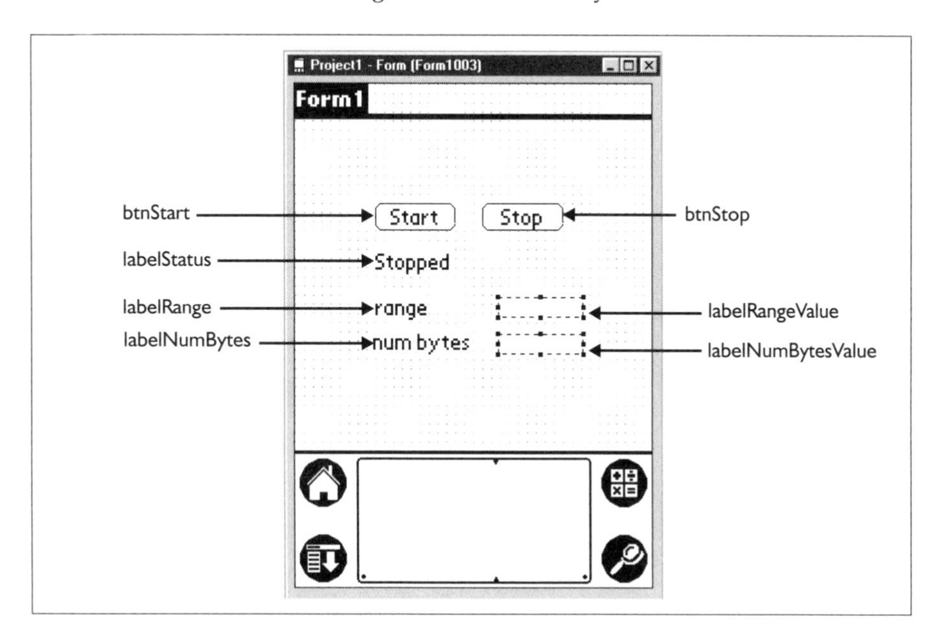

These two controls are named:

- ❑ LabelRangeValue
- ❑ LabelNumBytesValue

We will use these two labels to output information from the sensor reading function so you can get a feel for how to read sensor data from NS Basic.

As a general rule, you will want to open the serial port to the SV203 controller when your program starts up, and close the port when the program ends. To do this, right-click on the Project1 element in the Project Explorer window.

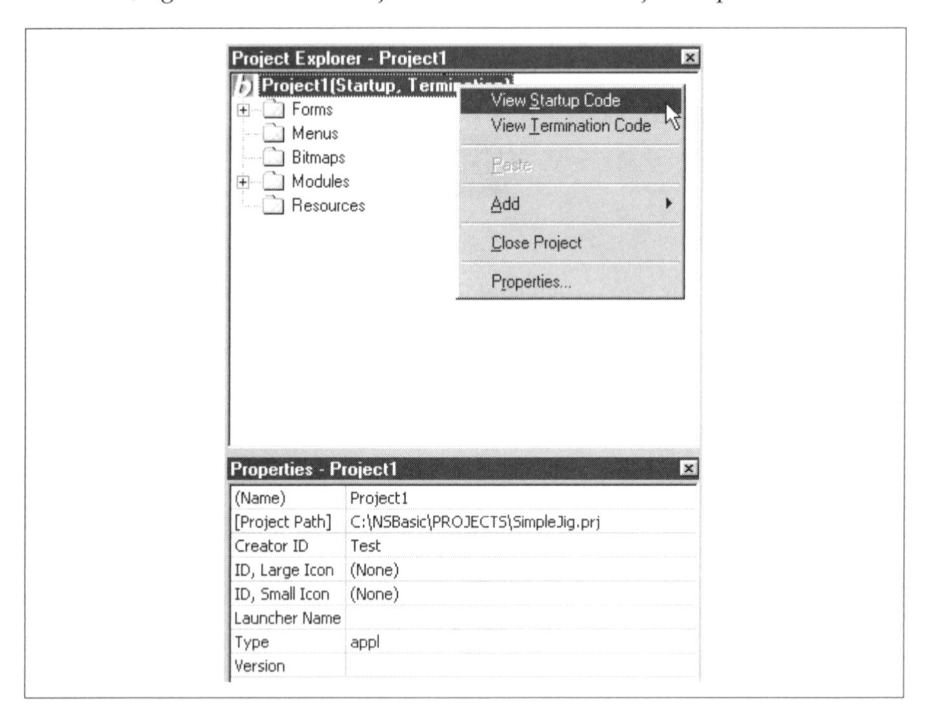

Select View Startup Code, and enter this code:

```
Sub Project_Startup()
    SerialOpen(0, 9600)
End Sub
```

This code will, as you might guess, open the serial port for you. Note that NS Basic automatically enters the Sub and End Sub lines of code, and all you need to enter is the SerialOpen( ) function call.

Now right-click the Project element again and select View Termination Code. Enter this code in the window:

```
Sub Project_Termination()
    SerialClose()
End Sub
```

Double-click on the Start button. Add the code (shown next) to the code window that appears. Note that your Sub line will probably use a different object ID, depending on when you added the button to the project.

```
Sub object1004()
    'start button
    lblStatus.label = "Running..."
    Call LoopMethod
End Sub
```

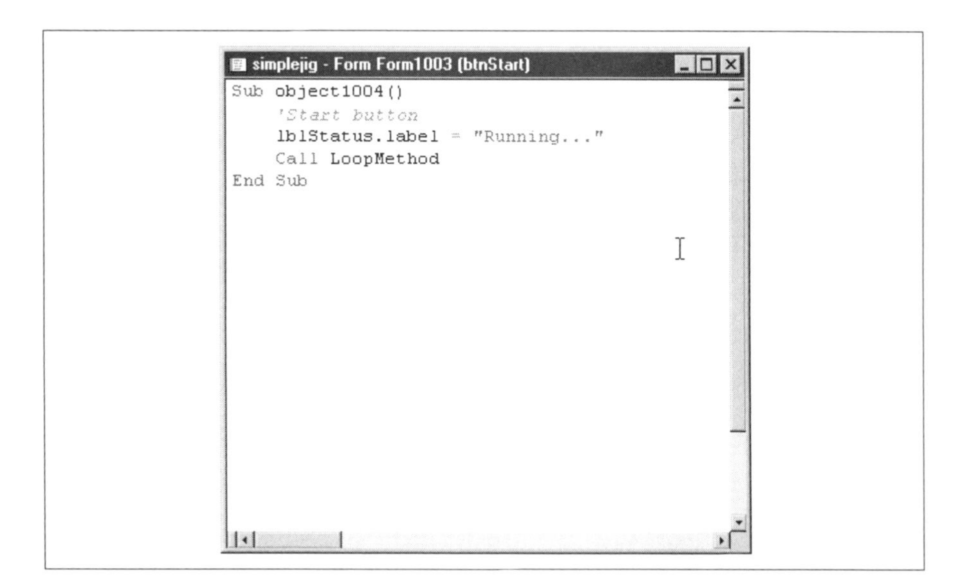

Double-click the Stop button and enter this code into the code window for the Stop button:

```
Sub object1005()
    'stop button
    done = 1
    lblStatus.label = "Stopped..."
End Sub
```

Right-click on the Modules folder in the Project Explorer and select Add New Module from the popup menu. Add the code shown in the next listing to the code module. You may need to edit the code so that it refers the correct objects for your project.

```
Global range as Integer
Global sv1 as String
Global sv2 as String
Global sv3 as String
Global done as Integer

Sub LoopMethod()
    done = 0
    Do while done <> 1
        Call ReadRange()
        If range > 55
            Call DoJig()
        End If
        Delay 1
        If SysEventAvailable()=1 Then
            done = 1
        End If
    Loop
    Return
```

```
End Sub

Sub DoJig()
    sv1 = "SV1M128"+chr(13)
    sv2 = "SV2M1"+chr(13)
    sv3 = "SV3M255"+chr(13)
    Call ServoControl()
    Delay 2

    sv1 = "SV1M255"+chr(13)
    sv2 = "SV2M255"+chr(13)
    sv3 = "SV3M255"+chr(13)
    Call ServoControl()
    Delay 3

    sv1 = "SV1M128"+chr(13)
    sv2 = "SV2M1"+chr(13)
    sv3 = "SV3M255"+chr(13)
    Call ServoControl()
    Delay 2

    sv1 = "SV1M1"+chr(13)
    sv2 = "SV2M1"+chr(13)
    sv3 = "SV3M1"+chr(13)
    Call ServoControl()
    Delay 3

    sv1 = "SV1M128"+chr(13)
    sv2 = "SV2M128"+chr(13)
    sv3 = "SV3M128"+chr(13)
    Call ServoControl()
    Return
End Sub

Sub ServoControl()
    Dim err as Integer
    err = serialsend(sv1,len(sv1))
    err = serialsend(sv2,len(sv2))
    err = serialsend(sv3,len(sv3))
    Return
End Sub
```

```
Sub ReadRange()
    Dim r as String
    Dim numbytes as Integer
    Dim readSensorStr as String
    Dim err as Integer

    range = -1
    readSensorStr = "BD1AD1"+chr(13)
    err = serialsend(readSensorStr, len(readSensorStr))
    numbytes = sysinfo(5)
    LabelNumBytesValue.label = str(numbytes)
    If (numbytes > 0) Then
        SerialReceive(r,numbytes,3.0)
        range = val(r)
        LabelRangeValue.label = str(range)
    EndIf
    Return
End Sub
```

The flow of this program is the same as it was before. When the Start button is tapped, the application runs a loop that reads the sensor, and does a little jig if it detects an object with sensor 1. When the Stop button is tapped, the program ends. Notice that while the functions to open, read, and write the serial port are different in NS Basic than in HotPaw Basic, the concept is still the same. The program opens the port, and reads and writes data to the SV203 as simple text strings. You can find more information on the serial port functions and other functions in the documentation for NS Basic.

Although we will not go through this code line by line, we will point out some of the functions we used. Within the LoopMethod( ) function, the call to the SysEventAvailable( ) function tells us whether the user has tapped a button or control. If this happens, the program assumes that the user wants to stop. Within the ServoControl( ) function, the serialsend(String, Integer) function sends a string to the serial port. This is the function that sends our commands to the SV203. Within the ReadRange( ) function, the sysinfo(5) function calls and returns the number of bytes waiting to be read from the serial port. The SerialReceive( ) function reads those bytes from the serial port.

Other Tools

HotPaw Basic and NS Basic are by no means the only tools available to you. There are several other BASIC systems for your Palm OS. The best place we know for finding programming tools for your PDA is www.PalmGearHQ.com. If you go there and search for "basic program," you will get a listing of additional tools and interpreters that you can play with.

Programming the PPRK with C

At the beginning of the chapter, we said that most C and C++ systems do not allow you to edit source code on the Palm. You need to edit, compile, and create the PRC file on your PC and then perform a hot sync operation to load the application to the PDA. In this section, we'll look at one exception to that.

PocketC

PocketC, available from www.orbworks.com, is a C compiler that allows you to create and compile C programs directly on your PDA. For the most part, though, as your C programs get bigger than just a few lines you'll want to edit your source code as memos in the Palm Desktop, and hot sync the memo to the PDA for compiling.

Download PocketC from the Web or from the CD-ROM, and unzip it onto your hard drive. The compiler is contained in the PocketC.prc file. If you are going to make use of any floating point math, you'll also need the mathlib.prc file. HotSync both of the files to your Palm and you're ready to go.

Find and tap the icon for PocketC in the Applications Launcher. This will bring up a program list form that shows all the compiled programs that PocketC knows about. If you just installed PocketC, this list will be empty. Below the form are three buttons. When programs are listed in the form, you can select the program and use the Execute button to execute the selected program. The Compile button displays a form that allows you to select and compile a program stored as a memo. The middle button, labeled Output, takes you to a form that displays any output created by a program.

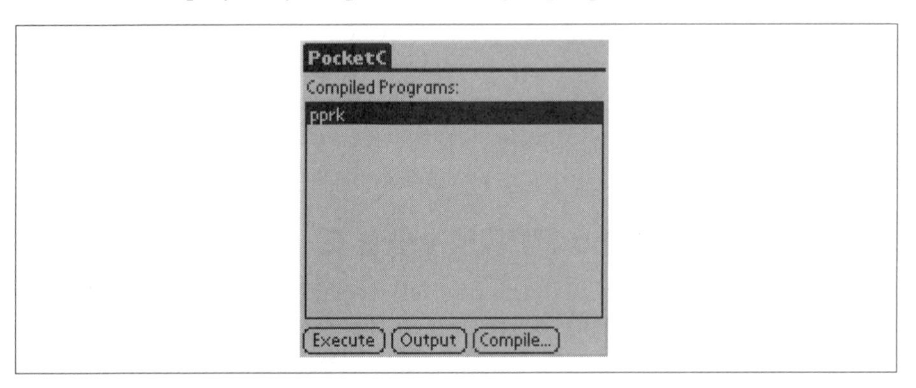

Hello, World!

Let's start with a simple example that will show you some of the features of PocketC. In this example, we'll draw a form on the screen, and use some buttons to accept input from the form. For now, don't worry about understanding every keyword or function in the listing. We just want to get a quick example working. This program is short enough that you could enter it directly into a memo on your PDA. However, most of the time it will be easier to type the memo into the Palm Desktop and then hot sync it to the PDA. (If you read the HotPaw Basic section above, you probably have a strange feeling of déjà vu about now.)

What do we want the program to do? First, it will display the phrase "Hello, World!" to the screen. Next, it will draw a button on the screen.

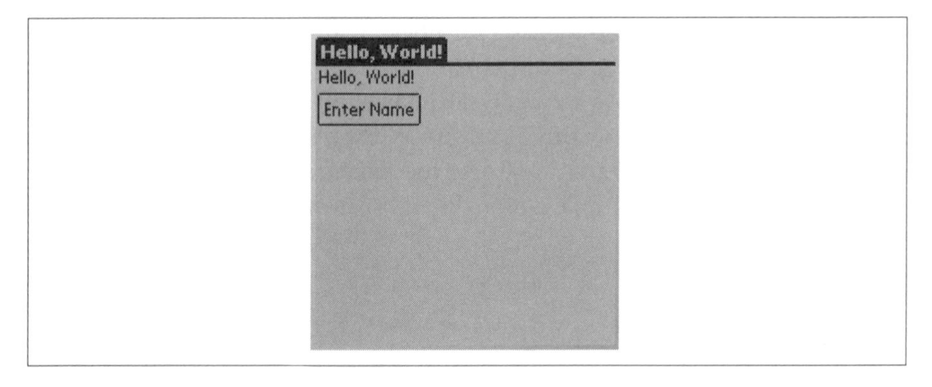

Every time the user taps the button, the program will display a message box that asks the user to enter a name.

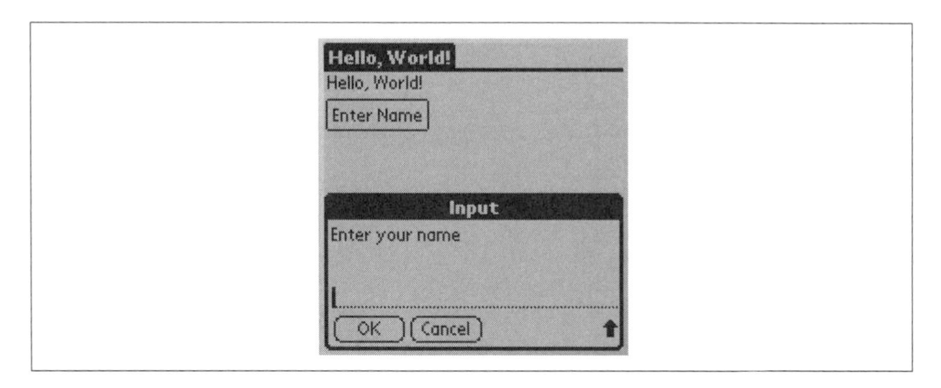

If the user enters a non-empty string, the program will change the displayed phrase to "Hello, *name*."

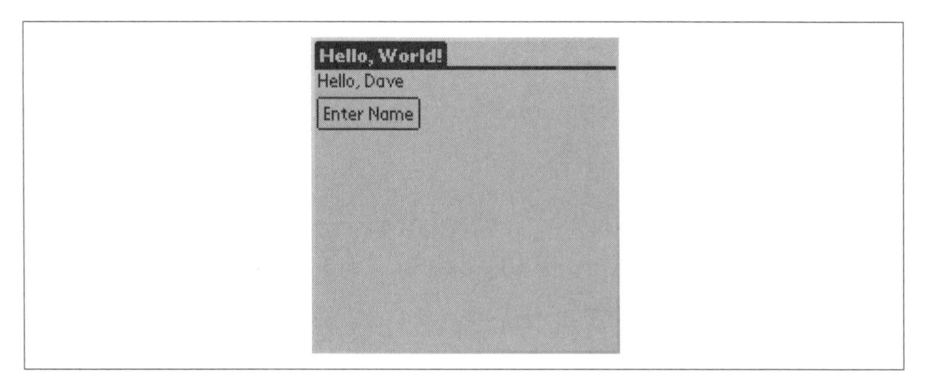

Here's our program:

```
//HelloWorld.c
main() {
  int x, y, ev;
  string mymsg, name;
  graph_on();
  mymsg="Hello, World!";
  title(mymsg);
  while(1) {
    clearg();
    text(2, 15, mymsg);
```

```
      frame(1, 2, 30, 55, 45, 2);
      text(5, 32, "Enter Name");
      waitp(1);
      x = penx();
      y = peny();
      if (x>2 && x<55 && y>30 && y<45) {
        name = gets("Enter your name");
        if (name != "") {
          mymsg = "Hello, " + name;
        }
      }
    }
  }
```

Let's go over this program quickly. Every PocketC program starts with the name of the program as a C++ style comment. The name must end in .c or .pc so that PocketC can recognize it as source code.

Every program must also have a single main( ) function. As with other C source code, the main( ) function is the starting point for the program. Also, you can have additional functions in your source code, either before or after the main( ) function.

Inside the main( ) function, the code declares some variables and then turns the graphics form on. The title( ) function changes the title of the graphics form. Then we go into the primary loop. Within the loop, the clearg( ) function clears the graphics form.

The code draws the message to the form using the text( ) function, and then creates a pseudo button using the frame( ) and text( ) functions. Don't worry that you need to do this for every program. There are several add-on libraries available at www.orbworks.com that provide functions for creating form controls. We only chose to do it this way for simplicity. If you plan to pursue PocketC programming, you should definitely check out the add-on libraries at OrbWorks.

The program waits for a pen event with the waitp( ) function. When this function returns, the x and y screen coordinates of the pen click can be obtained with the penx( ) and peny( ) functions. If the x and y coordinates are within the bounds of the pseudo button, the code displays a dialog box that asks the user to enter a name. If the name is non-empty, the message is changed. When the loop starts over, the screen is cleared and the current message is displayed.

Loading and Running Hello World

Start by creating the program as a memo on your Palm OS device. Alternately, you can create it as a memo in the Palm Desktop or a note in Outlook, and perform a hot sync operation to load the memo to the PDA.

After you have created the memo, launch PocketC on the PDA. Tap the Compile button to go to the compile screen. You will see a list of memos that contain programs; unless you've already entered other programs, this list will only show the Hello World program. Select the program and tap the Compile button to compile the program. If the program has an error, PocketC will display a dialog box telling you which line has a problem. When the program successfully compiles, PocketC will automatically return to the main screen. Select the compiled program and tap the Execute button to run the program.

PPRK Programming with PocketC

It's time to revisit the Simple Jig program again, this time using PocketC. If you don't want to type all this into a memo, we've included the source code in the \Other Applications\SimpleJig directory on the CD-ROM.

```
//SimpleJig.c

char cmd11[6] = {'S','V','1','M','1'};
char cmd1128[8] = {'S','V','1','M','1','2','8'};
char cmd1255[8] = {'S','V','1','M','2','5','5'};

char cmd21[6] = {'S','V','2','M','1'};
char cmd2128[8] = {'S','V','2','M','1','2','8'};
char cmd2255[8] = {'S','V','2','M','2','5','5'};

char cmd31[6] = {'S','V','3','M','1'};
char cmd3128[8] = {'S','V','3','M','1','2','8'};
char cmd3255[8] = {'S','V','3','M','2','5','5'};

char range[10];
int rv;

initCmds() {
  cmd11[5] = (char)13;
  cmd1128[7] = (char)13;
```

```
  cmd1255[7] = (char)13;
  cmd21[5] = (char)13;
  cmd2128[7] = (char)13;
  cmd2255[7] = (char)13;
  cmd31[5] = (char)13;
  cmd3128[7] = (char)13;
  cmd3255[7] = (char)13;
}

readSensor() {
  char c[8];
  int x;
  strtoc("BD1AD1"+(char)13, c);
  c[7] = (char)0;
  puts("Entered readSensor\n");
  sersenda(c, 7);
  rv = 0;
  rv = serdata();
  puts("rv=" + format(rv,0)+"\n");
  if (rv > 0) {
    serrecva(range, rv);
    range[rv] = (char)0;
    x = ctostr(range);
    puts("range=[" + format(x,0)+"]\n");
    text(40,85,format(x,0));
  } else {
    puts("No data on serial port\n");
  }
}

doJig() {
  sersenda(cmd1128,8);
  sersenda(cmd21,6);
  sersenda(cmd3255,8);
  sleep(2000);

  sersenda(cmd1255,8);
  sersenda(cmd2255,8);
  sersenda(cmd3255,8);
  sleep(3000);

  sersenda(cmd1128,8);
```

```
  sersenda(cmd21,6);
  sersenda(cmd3255,8);
  sleep(2000);

  sersenda(cmd11,6);
  sersenda(cmd21,6);
  sersenda(cmd31,6);
  sleep(2000);

  sersenda(cmd1128,8);
  sersenda(cmd2128,8);
  sersenda(cmd3128,8);
}

main() {
  int x, y, done, ev, rVal;
  string rangeStr;
  initCmds();
  graph_on();
  title("Simple Jig");
  clearg();
  while(1) {
    seropen(9600, "8N1N", 10);
    frame(1, 40, 50, 75, 65, 2);
    text(43, 52, "Start");
    frame(1, 85, 50, 120, 65, 2);
    text(88, 52, "Stop");
    text(40, 70, "Stopped   ");
    x = -1;
    y = -1;
    while(x == -1 && y == -1) {
      waitp();
      x = penx();
      y = peny();
      if(x<40 || x>75 || y<50 || y>65) {
        x = -1;
        y = -1;
      }
    }
    text(40, 70, "Running   ");
    done = 0;
    while(!done) {
```

```
        readSensor();
        if (rv > 0) {
          rangeStr = ctostr(range);
          rVal = rangeStr; //converts string to int
          if (rVal > 55) doJig();
        }
        sleep(1000);
        ev = event(0);
        if (ev == 3) {
          x = penx();
          y = peny();
          if (x>85 && x<120 && y>50 && y<65) done = 1;
        }
      }
      text(40, 70, "Stopping");
      serclose();
    }
  }
```

Simple Jig Explained

The program starts by declaring a set of character arrays. These character arrays hold the servo commands that will be sent to the BrainStem. We named the character arrays using the string cmd followed by the servo number and the servo position. So, looking above, you can see that the arrays that hold the commands to set servo 1 to positions 1, 128, and 255 are cmd11, cmd1128, and cmd1255. The arrays for the commands for servos 2 and 3 are named similarly. We also declare two other variables at the top of the program. The variable range[10] is a character array that will hold the sensor readings from the SV203. The variable rv will hold the number of characters waiting on the serial port.

You may have noticed that the last character in the command arrays was not set when the character array was declared. The final character of each command needs to be the ASCII code 13. We use the initCmds( ) function to set the final character of the array. This function will be one of the first things called by the main( ) function.

The next function is the readSensor( ) function. This function contains the code to get a sensor reading from the SV203 controller. The code starts by declaring an eight-character array named c to hold the command to read a sensor. It also declares a temporary variable named x that will be used for some debugging output. The code calls the strtoc( ) function to put the sensor reading command (BD1AD1) into the character array. The code then sets the last character of the array to be \0. The puts( ) function call sends a string to an output screen managed by PocketC. You can see the output by tapping the Output button on the main PocketC screen. The sersenda( ) function outputs a character array to the serial port. The serial port must already be open for this command to work; we will see the command to open the serial port in the main( ) function. The arguments to the sersenda( ) function are the character array and the number of characters to send. The serdata( ) function returns the number of bytes waiting to be read on the serial port. If there is data on the serial port, the code reads it with a call to serrecva( ). The arguments for this function are the character array to store the data in, and the number of bytes to read. To ensure that the character array can be converted to a string, the code sets the array element following the last byte to 0. The code then calls the ctostr( ) function. This converts the character array to a string and assigns the result to our int variable x. Unlike standard C, PocketC allows you to assign a string to an int. If the string starts with a number, the number is assigned to the variable. Thus, you don't need a special function to convert a string to an int. The value of x is then sent to the Output display and to the screen of the PDA.

Following the readSensor( ) function is the doJig( ) function. This function uses the sersenda( ) function to send the command to the serial port. The serial port must already be open for sersenda( ) to work. The sleep( ) function stops the program for the given number of milliseconds. One thousand milliseconds is one second.

The main( ) function is what PocketC calls to start the program. The program starts by declaring some variables, calling the initCmds( ) function, and setting up the display. The next important function call is the call to open the serial port:

```
seropen(9600, "8N1N", 10);
```

As seen in the code snippet above, the function seropen( ) opens the serial port. The parameters set the serial port to 9600 baud, 8 data bits, no parity, 1 stop bit, and no flow control. The final parameter (10) is a timeout value in hundredths of a second.

After opening the serial port, the main( ) function draws two buttons on the screen for starting and stopping the program. When the user taps the Start button, the program starts a loop that calls the readSensor( ) function, and calls the doJig( ) function when sensor 1 detects an object. When you click the Stop button, the program closes the serial port using the serclose( ) function.

Loading and Running SimpleJig.c

Start by creating the program as a memo on your Palm OS device. Alternately, you can create it as a memo in the Palm Desktop or a note in Outlook, and perform a hot sync operation to load the memo to the PDA.

After you have created the memo, launch PocketC on the PDA. Tap the Compile button to go to the compile screen. You will see a list of memos that contain programs. Select the SimpleJig.c program and tap the Compile button to compile the program. If the program has an error, PocketC will display a dialog box telling you which line has a problem. When the program successfully compiles, PocketC will automatically return to the main screen.

Connect the PDA to the robot and turn on the robot. Select the SimpleJig program in the main PocketC screen and tap the Execute button to run the program. Put the robot on the floor and place an object in front of sensor 1. When the sensor detects the object, it will display the range reading on the screen. If the object is close enough, the robot will perform its jig. When you've seen enough, tap the Stop button to stop the program. Go back to the PocketC main screen, select SimpleJig, and tap the Output button. This will show you the debug output from the puts( ) function calls.

The Robot Geek Says

All of the software tools for creating, programming, and running programs can be found on the CD-ROM for this book. They are all demo or trial versions of the software. If you want to purchase the full version of any package, or you want to search for the latest versions and other information, you should go to the web sites of these products:

- **HotPaw Basic** www.hotpaw.com/hotpaw
- **MathLib** www.radiks.net/~rhuebner/mathlib.html
- **NS Basic** www.nsbasic.com/palm
- **PocketC** www.orbworks.com/pcpalm
- **PocketPPRK** home.townisp.com/~kmcentire/palm/pprk/pprk.htm
- **Pocket Toolbox** www.geocities.com/retro_01775/PToolboxLib.htm

We would like to thank the following people or companies for allowing us to include their software with this book:

- Ron Nicholson, creator of HotPaw Basic
- Rick Huebner, creator of MathLib
- NS Basic Corporation
- OrbWorks
- Karl McEntire, developer of the PocketPPRK application
- Joseph Stadolnik, developer of Pocket Toolbox
- Acroname Corporation, for the TEA language and SDKs
- Carnegie Mellon, PPRK project, for the original PPRK software

Chapter 8

Taking Control of the BrainStem Robot

In much of this book, we've been covering both the original PPRK built with the Pontech SV203 controller and the PPRK built with the BrainStem controller. This is the second of two chapters that focus on just one of the versions. In the previous chapter, we explored various languages for programming the SV203. In this chapter, we will focus exclusively on the BrainStem controller.

We will first look at writing programs in a language named TEA (Tiny Embedded Applications). Using TEA, you can write programs for the BrainStem that look very much like programs written in the C language. This is not surprising, because TEA is a subset of C. So, if you're a C programmer, you should find TEA simple to use. You can also use TEA to write programs in a different mode. Acroname calls this mode "Reflex," and as you will see, it provides a powerful toolset for writing programs that can react to stimuli.

Finally, we'll look at writing Java programs for the BrainStem. Java is one of the major programming languages today. We will see how to use Java together with some Java classes developed by Acroname to write programs for the BrainStem controller.

The Robot Geek Says

So You Want to be a BrainStem Hacker

In this chapter, we assume that you already know how to program in at least one computer language. This chapter will not be a tutorial on how to program, but rather how to use what you already know to program for the SV203 or BrainStem controller. If you want to learn how to program, there are many other books you can try. Here are a few suggestions:

- *Java 2: A Beginner's Guide*, by Herb Schildt, from McGraw-Hill/Osborne
- *Teach Yourself Java*, by Joseph O'Neil, from McGraw-Hill/Osborne
- *Thinking In Java*, by Bruce Eckel, from Prentice Hall
- *Beginning J2EE 1.4*, by Kevin Mukhar, from Wrox Press (No robot information whatsoever, but lots of Java programming. Since we plugged one of Dave's books in the previous chapter, this time it's Kevin's turn.)

Programming the BrainStem Using TEA

If you have a PPRK that uses the BrainStem controller, your easiest route to programming is to use the TEA (Tiny Embedded Applications) language. TEA is a subset of the C programming language, so if you know C or Java, you should be able to easily program in TEA.

BrainStem Modes

The BrainStem can actually operate in a number of different modes. These modes are called Slave mode, Reflex mode, and TEA mode.

In Slave mode, the robot is completely controlled through a host device. This host device can be a PC or a PDA. The host receives sensor readings from the BrainStem, performs calculations, and sends servo commands to the BrainStem. All the computation for controlling the robot occurs on the host, and the BrainStem module simply acts as a conduit for data and commands.

In Reflex mode, you create a number of small code blocks, called reflexes and messages, that are stored on the BrainStem controller. Reflexes consist of a list of references to messages; messages consist of some BrainStem command that you want to be executed. Events, such as a sensor reading or another reflex, cause one or more reflexes to be initiated. The reflex causes the commands in its list to be executed. You can see that these are analogous to the reflexes of your own body. Tap your knee in the right spot and your lower leg reflexively reacts. Sneeze and your eyelids close.

Finally, there is TEA mode. This is kind of a mixture of Slave and Reflex modes. Similar to Reflex mode, you store small programs on the BrainStem controller. However, unlike Reflex mode, TEA programs don't execute in response to events or other TEA programs. Some host must command the BrainStem controller to execute the TEA programs. However, after the host has sent the execute command, the BrainStem can be disconnected from the host and all processing occurs on the BrainStem controller.

We will be looking primarily at TEA mode in this section. Later we will take a brief look at Reflex mode. The final section in this chapter will look at using Java in Slave mode.

Installing Acroname Software

To use TEA mode, you will need the Console application and a library of TEA functions and subroutines, both from Acroname. We've included all the files you need on the CD-ROM for the book. But if you want to make sure you have the latest and greatest files, you can get them at www.acroname.com. The file you need is the pprk_sdk.zip or pprk_sdk.sit file. Acroname has created different versions of the software for many different systems. You will need to download the software for your system (Windows PC, Mac, or Palm OS). Unpack the file to your hard drive.

In the directory structure that is created when you unpack the pprk_sdk archive file are the directories aBinary, aSystem, and aUser. The aBinary directory contains programs that help you use your BrainStem controller. One of these is a program called Console, which is used to load programs from a host to the BrainStem. Another program is Config, which allows you to check the operation of the sensors and servos. The aSystem directory contains the library files that you can use to create BrainStem programs. The aUser directory is where you will place your TEA source code.

If you are using a Palm PDA as a host, TEA source code files should be placed in the MemoPad. You'll find it much easier to edit the files on a PC, and then perform a HotSync operation to put the memos on your PDA. You can edit TEA files on the PDA if you want, but it is a cumbersome process.

Writing TEA Programs

Since TEA is based on C, if you know C (like I know C), you won't have any problems at all. The big difference between the two is that TEA does not use pointers. Another difference between TEA and C is with a right shift operation. In TEA, a right shift on a signed or unsigned variable will perform an arithmetic shift. This means the sign bit is always shifted right into the value. In Ansi C, a right shift on an unsigned variable shifts zeros into the value. A final difference is that although you can include files in your TEA program, as we will see shortly, you don't include header files. TEA programs do not use header files like C or C++; the files that you include are other source files.

The main entry point for all TEA programs is the main( ) function. In this book, we will be using a form of main( ) that looks like this:

```
void main()
```

Recall that this means the main( ) function takes no arguments and returns no values. However, in TEA the main( ) function can also have a return type of char or int, and it can take an arbitrary number of parameters. Within main( ), you can execute program logic and call other functions or subroutines. These functions and subroutines can be in the same file, or they can be included from other files.

Just as with C or C++, you can include other source files using the #include macro:

```
#include <aPPRK.tea>
```

The command shown above includes the named source file, aPPRK.tea, into the current source file. Recall that with TEA programs, the file you include, aPPRK.tea for example, is a source code file, and not a header file. Code in the current source file can call any of the functions or subroutines defined in the included file.

Subroutines and functions are code modules that can be called just like functions and subroutines in C. The primary difference between the functions and subroutines is that functions return a value and subroutines do not.

```
int myFunction() /* returns an int */
void mySubroutine() /* returns nothing (void) */
```

A Sample TEA Program

If you went through Chapter 5, you have already encountered one TEA program, simple.tea. This is the program that waits until sensor 1 detects an object, and then it does a little jig that consists of moving away from the object, turning one direction, then reversing the turn, and finally moving forward.

Rather than go through that code again, let's examine another of the TEA sample programs and look a little more closely at how it works. After we do that, you may want to investigate the other sample programs on your own or try writing a TEA program yourself.

The program we will look at is the one named chase.tea in the aUser directory. Here is the complete source code for the program:

```
/* chase.tea                                                    */
/* BrainStem PPRK program                                       */
/* Move toward sensor detection, otherwise spin                 */
/* Only moves in direction of one sensor                        */
#include <aCore.tea>
#include <aPPRK.tea>
#include <aA2D.tea>

#define     SPEEDR     100
#define     SPEEDL     100
#define     DRANGE     175

void main()
{
  int b;
  int r1;
  int r2;
  int r3;

  /* five second wait before starting */
  aCore_Sleep(30000);
  aCore_Sleep(20000);

  while (1)
  {
    r1=aA2D_ReadInt(APPRK_IR1);
    r2=aA2D_ReadInt(APPRK_IR2);
    r3=aA2D_ReadInt(APPRK_IR3);

    b=0;
    if (r1>DRANGE) b = b | 1;
    if (r2>DRANGE) b = b | 2;
    if (r3>DRANGE) b = b | 4;

    switch (b)
    {
      case 1:
        aPPRK_Go(0,SPEEDL,-SPEEDL,0);
        b=1;
```

```
        break;
      case 2:
        aPPRK_Go(-SPEEDL,0,SPEEDL,0);
        b=1;
        break;
      case 4:
        aPPRK_Go(SPEEDL,-SPEEDL,0,0);
        b=1;
        break;
      default:
        b=0;
        break;
    }

    if (b==0)
    {
      aPPRK_Go(0,0,0,SPEEDR);
    }
    else
    {
      aCore_Sleep(5000);
      aPPRK_Go(0,0,0,0);
      aCore_Sleep(5000);
    }
  }
}
```

chase.tea Line by Line

Let's spend a few minutes examining the chase.tea program to get an understanding of the kinds of things that go into a TEA program. The program begins with four comment lines that start with the following comment:

```
/* chase.tea                                              */
```

TEA follows ANSI C, so the markers /**/ above indicate that what appears between the markers is a comment and is ignored by the compiler. Within TEA code you can also use // to indicate comments like this:

```
//chase.tea
```

Following the comment lines in the source file are some include statements:

```
#include <aCore.tea>
#include <aPPRK.tea>
#include <aA2D.tea>
```

These three files are part of the library that comes with the Acroname software. The aCore file contains functions and subroutines for sending data to and from the ports, and it contains a subroutine, aCore_Sleep( ), that pauses the program. The aPPRK file contains a single routine for commanding the servos. The file aA2D contains two functions that read data from the analog ports of the BrainStem. Finally, both aCore and aA2D include the file aIOPorts, which defines constants that can be used in the TEA programs. The include statements are followed by some constant definitions:

```
#define     SPEEDR      100
#define     SPEEDL      100
#define     DRANGE      175
```

These lines define three constants that will be used in the code that follows. Using constants is better than using the literal numbers in your code. It makes the code easier to understand because anyone reading the code can make a good guess about what `SPEEDR` refers to, but guessing what 100 refers to would be much more difficult.

Next, we get to the starting point for the program. Each program you create and run must have one and only one main( ) function. The code for chase.tea uses void main( ), but recall that main( ) can return char or int, and take parameters. The main( ) function is the function that the TEA VM calls to begin execution of your program.

```
void main()
{
```

Next, the program defines a number of variables:

```
  int b;
  int r1;
```

```
int r2;
int r3;
```

The variable b tracks which sensor detected an object. The variables r1, r2, and r3 hold the readings from sensors 1, 2, and 3.

This is followed by two sleep commands. The purpose of the aCore_Sleep(int) function is to pause the program. The time that the program pauses is given by the int parameter. This parameter specifies how many ten-thousandths of a second that the program pauses. So, to pause for 1 second, you pass the value 10,000 to the subroutine. In the code, the program waits for 3 seconds, then 2 seconds, for a total wait time of 5 seconds.

```
/* five second wait before starting */
aCore_Sleep(30000);
aCore_Sleep(20000);
```

The Robot Geek Says

The Big Sleep

You might be wondering why the chase.tea source code uses two calls to the aCore_Sleep(int) function, rather than:

```
aCore_Sleep(50000);
```

The reason is that TEA uses both signed and unsigned int parameters. The parameter for the sleep function is defined to be an int. The range for signed int parameters is –32768 to 32767. Thus, the largest number you can use for aCore_Sleep is 32767. If you need the robot to sleep for more than 3.2767 seconds, you will need to use multiple function calls.

TEA uses the following data types:

- **void** The void type represents no data. It is typically used to show explicitly that there is no return value from a routine or no parameters for a routine.
- **char** The char type represents a signed byte. It has a range of –128 to 127.

- **unsigned char** The unsigned char type represents an unsigned byte. It has a range of 0 to 255.
- **int** The int type represents a signed integer comprised of 2 bytes. It has a range of –32768 to 32767.
- **unsigned int** The unsigned int type represents an unsigned integer comprised of 2 bytes. It has a range of 0 to 65535.
- **string** The string type is a fixed series of ASCII-encoded text characters.

Now we get to the main loop of the program. The first thing that occurs in the loop is that the program calls the aA2D_ReadInt( ) function for each of the three sensors. You can see that the function takes an argument that tells the function which sensor to read. In the chase.tea program, the code uses symbolic constants to refer to each of the sensors. The symbolic constants APPRK_IR1, APPRK_IR2, and APPRK_IR3 are defined in the aPPRK.tea file.

```
while (1)
{
  r1=aA2D_ReadInt(APPRK_IR1);
  r2=aA2D_ReadInt(APPRK_IR2);
  r3=aA2D_ReadInt(APPRK_IR3);
```

This is followed by code that sets the value of the variable b based on which sensors detect an object. If one or more of the sensors detect an object, the value of b will be OR'd with 1, 2, or 4, depending on which sensors detect an object. If only a single sensor detects an object, b will have the value 1, 2, or 4. If more than one sensor detects an object, b will have the value 3, 5, 6, or 7. If no sensor detects an object, none of the if statements will evaluate to true and b will remain 0.

```
  b=0;
  if (r1>DRANGE) b = b | 1;
  if (r2>DRANGE) b = b | 2;
  if (r3>DRANGE) b = b | 4;
```

The code then executes a switch statement based on the value of b. If b is 1, 2, or 4, the program commands the servos so that the robot moves toward the detected object. If b is any other value, the default block executes and sets b to 0.

```
      switch (b)
      {
        case 1:
          aPPRK_Go(0,SPEEDL,-SPEEDL,0);
          b=1;
          break;
        case 2:
          aPPRK_Go(-SPEEDL,0,SPEEDL,0);
          b=1;
          break;
        case 4:
          aPPRK_Go(SPEEDL,-SPEEDL,0,0);
          b=1;
          break;
        default:
          b=0;
          break;
      }
```

Finally, as seen in the following code, if b is 0 (meaning no sensor detected an object, or more than one sensor detected an object), the servos are commanded to rotate the robot. Otherwise, only a single sensor detected an object, and the else block waits for half a second, stops the servos, and waits for another half second. From there, the program jumps back to the start of the loop.

```
      if (b==0)
      {
        aPPRK_Go(0,0,0,SPEEDR);
      }
      else
      {
        aCore_Sleep(5000);
        aPPRK_Go(0,0,0,0);
        aCore_Sleep(5000);
      }
    }
  }
```

In summary, this program causes the robot to spin around until it sees an object with one sensor. It then moves toward that object. If it loses the object, or if it detects objects with more than one sensor, it reverts to spinning again.

Steeping a Steaming Cup of TEA

Before the TEA program above can be loaded and executed on the BrainStem module, it must be compiled into a form that is executable by the BrainStem. Unlike C code, TEA code is not compiled into some executable machine language form. TEA code is compiled into a form that can be executed by a TEA virtual machine.

The Robot Geek Says

Virtual Machines

- A virtual machine is a computer implemented as a software program that runs on some other host computer.
- A virtual machine contains code specific to a processor and operating system.
- Virtual machines allow you to use the same executable on multiple platforms, as long as that platform has a virtual machine. (Compare this to a C program written and compiled for the Palm OS, for example. That application cannot run on a Pocket PC device.)
- Java is another language that runs on a virtual machine.

The Console program is used to compile, load, and launch TEA programs. If you are running a PC, you should find the Console program in the aBinary directory.

If you are using a PDA, you need to perform a HotSync operation to load the console.prc application and supporting Palm database files to the PDA. Navigate to the \aBinary directory of the Acroname software. Add console.prc and all the PDB files in the \aBinary directory to the install file list, and then perform a HotSync operation to load them to the PDA. The current build of the console program when we wrote this chapter included these files:

- console.prc
- aIO.pdb
- aLeaf.pdb
- aSteep.pdb
- aSteepGen.pdb
- aSteepOpt.pdb

- aStem.pdb
- aTEAvm.pdb
- aUI.pdb

If you are using a later version of the Acroname software, the list of files you find may be different. The key point is to load the console.prc and all the .pdb files to your Palm OS device.

You will also need to enter the TEA files as memos. The easiest way to do this is to create the memos in the Palm desktop and then perform a hot sync operation to load them to the PDA. You will need to edit the TEA files slightly. The first line to TEA file memos should be the name of the file. For example, the first line of the chase.tea memo file should be "chase.tea" without any other characters or comment markers. You also need memo files for each of the included files. Thus, you need the following files as memos on your PDA:

- chase.tea
- aA2D.tea
- aPPRK.tea
- aCore.tea
- aIOPorts.tea

Additionally, you may need to edit some of the TEA files by removing some comment lines. We found that a couple of the TEA files were too large to fit in a memo without editing them. (Memo pad files have a limit of 4,096 characters.) Finally, launch the Console program. You are now ready to compile a program.

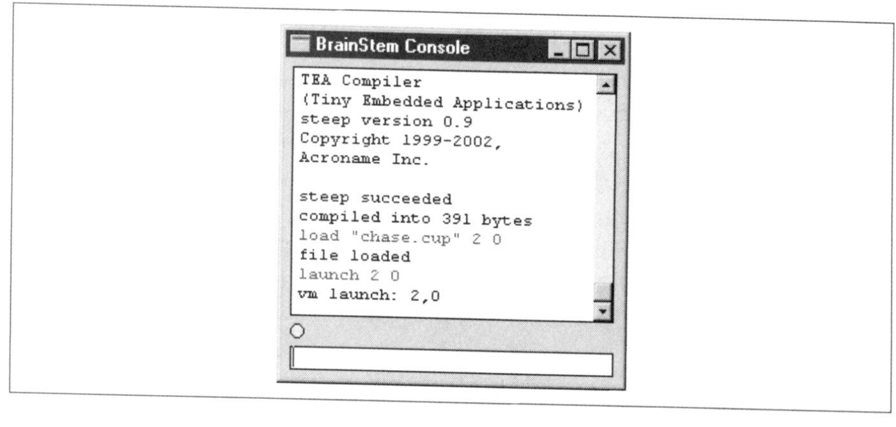

TEA code is compiled using the command steep followed by the quoted program name:

```
steep "chase.tea"
```

This will take the TEA source code and compile it into a form that can be used by the virtual machine on the BrainStem controller. If you are running on a PC, the compiled file will be placed into the \aObject directory. The name of the file will be the same, but with a new extension: .cup. Thus, you now have a cup of TEA. If you are running on a Palm, the cup file will be stored as another memo.

The Robot Geek Says

Examining a cup File

Although you can look at a cup file in its native format, it won't do you much good since you don't speak or read in byte code. However, you can process the byte code in the cup file into a somewhat more readable format. To create a readable file out of byte code in a cup file, you can

- Type **ast simple** on the console command line. This creates a file named simple.ast that shows the annotated syntax tree of the program.
- Type **dsm simple** on the console command line. This creates a file named simple.dsm that contains the assembler style machine code that corresponds to the byte code.

Downloading and Launching Programs

After you successfully compile the TEA program, you need to load it to the BrainStem controller and launch the program. This is accomplished using the load and launch commands. Launch the Console application. With the BrainStem connected to the host, switch the robot power on. If the BrainStem connects to the Console, you should see a green light on the BrainStem start blinking. A similar light will start blinking in the Console application. If the programs do not connect, you can try turning the BrainStem on first, then launching the Console application. Now execute the load command:

```
load "simple.cup" 2 0
```

The first parameter to the load command is the name of the program. This is followed by the BrainStem ID. Each BrainStem controller has a board identification number. This allows a single host to command multiple controllers at the same time. With multiple BrainStems, each BrainStem is given a different ID, and the host addresses each controller by its ID. The default ID for the BrainStem is 2, and since we have not changed the ID, that is the argument we use in the load command. Finally, each BrainStem has 10 slots, numbered 0 to 9, in which you can store TEA programs. The slot number is the final parameter.

Manually Launching Programs

Since the BrainStem has 10 slots, you can store up to 10 programs on the BrainStem at the same time. In addition, you can have multiple programs running at the same time. To command the BrainStem to execute one or more programs, you use the launch command:

```
launch 2 0
```

The first parameter is the board ID, and the second is the slot number of the program to execute. The default board ID for BrainStem controllers is 2. The slot number can be any number from 0 to 9. If you want to run more programs at the same time, you would issue another launch command using the slot numbers of the other programs.

At this point, the robot should begin executing the program. If it detects an object on one of its sensors, it will move toward that object. If no sensor detects an object, or more than one sensor detects an object, it will stop moving and start spinning.

Notice that chase.tea, simple.tea, and each of the other sample programs that come with the Acroname software has a five-second delay at the start of each program. This allows you time to disconnect the PPRK from the host and set it down on the floor. If you program the BrainStem to auto-execute a program, it also provides time for the BrainStem to finish its initialization sequence before executing the program. We will see how to do this in the next section.

Auto-launching Programs

While you are experimenting with different programs, you will want the flexibility of having a host computer connected to the robot, and manually launching the program or programs you are working on. However, once you have a working program, you may want the program to be launched automatically every time you turn the robot on. With auto-launching, you can take the robot anywhere and be free from the need to have a host computer to operate the robot.

After you have steeped and loaded a TEA program onto the BrainStem, you can instruct the BrainStem to automatically launch that program by issuing two commands. The two commands will set the BrainStem to automatically launch a loaded program the next time that the BrainStem boots up:

- ❑ cmdVAL_SET
- ❑ cmdVAL_SAV

These two commands are defined in the BrainStem command reference files. You can find the reference files on the CD-ROM, or you can access them at Acroname's web site. Information on the commands is in the Commands section of the reference; information on bootstrapping programs is in the TEAvm section of the reference.

The cmdVAL_SET command is used to set a system variable. In this case, we will set the variable that tells the BrainStem to auto-launch a program. The cmdVAL_SAV command saves the variable setting in the EEPROM so it is available the next time the BrainStem boots up.

When issuing BrainStem commands to the BrainStem, the general syntax is

```
module_id command_id [parameter [,parameter]]
```

As we discussed above, the default module ID for BrainStem controllers is 2. The command ID for cmdVAL_SET is 18, and cmdVAL_SAV is 19 (you can find this in the Commands section of the reference). cmdVAL_SET takes two parameters: the index of the system variable and a new value for the variable. You can set up to four boot programs using system variables 15, 16, 17, or 18. The new value will be the program slot of the program to be launched. So to set and save a variable to command the program in slot 0 to auto-launch, type the following two lines into the Console program, pressing Enter after each line to send each line:

```
2 18 15 0
2 19
```

To disable the auto-launch, you simply need to pass an invalid slot ID for the final parameter:

```
2 18 15 255
2 19
```

One final note: when you create programs that you plan to auto-launch, the first thing the program should do is go to sleep for a few seconds. This allows the BrainStem to complete its bootup and initialization before the auto-launched program continues executing. Recall from the chase.tea and simple.tea programs that we've looked at that Acroname uses a five-second delay.

Writing Your Own TEA Programs

At this point, you should be ready to begin writing and executing your own TEA programs on the BrainStem. There are many useful functions contained in the library files in the aSystem directory. But for the most part, you only need a few things to get started with TEA:

- The TEA file that is the starting point for your program must have a main( ) function.
- You can read sensors with the aA2D_ReadInt(char) function. Use one of the predefined constants APPRK_IR1, APPRK_IR2, or APPRK_IR3 as arguments to this function.
- You can command the servos with the aPPRK_Go(char,char,char,char) function. Note that the arguments are defined as chars to force you to pass arguments in the range 0 to 255.
- You can pause the program with the aCore_Sleep(int) method.

Program the BrainStem Using Reflexes

Our whole point in writing this chapter has been to show you how to use what you already know to program your robot. This section is going to violate that purpose. While using reflexes isn't that difficult, it is not something you probably already know how to do. It consists of new concepts and a new programming language that you would need to learn. For those reasons, we will not spend a great deal of time on reflexes. If, after reading this section, you want to pursue reflexes in more detail, you should consult the reference manual provided with the Acroname software.

How Reflexes Work

In a word: pretty darn simply. Okay, that's really three words. Not two. Not four. Three. No more. No less. Three being the number of words and the number of words being three.

Reflex mode uses messages and reflex vectors. A message is a specific piece of code stored on the BrainStem controller. This message consists of a BrainStem command and the parameters needed by the command. A reflex vector is simply a list of "pointers" to messages. When a reflex is triggered in response to an input, the messages in the reflex vector are executed. Among the numerous actions that messages can perform are reading sensors, commanding servos, and calling other reflexes.

Both the messages and the reflex vectors are compiled and downloaded to the BrainStem using the Console application we looked at in the previous section. After the messages and reflex vectors are downloaded to the BrainStem, they will execute in response to specific stimuli.

A Simple Reflex Program

In this section, we'll construct a very simple reflex program, and show you how to compile and download it using the Console application. Programming a reflex requires you to think a little differently. With TEA, C, and BASIC, one tends to think linearly: the program starts here and proceeds through a series of commands. With C++ and Java, one should be thinking in objects: the program consists of components that cooperate by sending messages (method calls) to each other. But in Reflex mode, you need to think about stimuli and responses. There is not a controlling main( ) method to coordinate actions. If a stimulus occurs, a reflex may or may not be activated. Stimuli can occur in any order with no beginning or end. Once a response is complete, the reflex stops executing until it receives another stimulus. As we mentioned above, messages can cause other reflexes to execute, so it is possible to create stimulus-response cycles with messages and reflex vectors.

Let's do something simple. We'll have the robot simply move back and forth continuously. So what are the commands we need?

- ❑ Command servos to move forward
- ❑ Countdown two seconds
- ❑ Command servos to reverse

- ❑ Countdown two seconds
- ❑ Repeat

Let's start with the first message. In a reflex source file, commands are contained in block of code that looks like this:

```
message[index] {
     module, command, parameters
}
```

Each message has an index so that we can reference messages by index number. The message body consists of the BrainStem module ID (which is 2), a BrainStem command, and parameters needed by the command. So, for example, consulting the BrainStem reference manual that comes with the software, you can find the command cmdSRV_REL for commanding a servo. This command takes three parameters. The reference manual states that the first parameter is the servo number (0–3), the second parameter is the servo position, and the third parameter is the relative direction for the servo to rotate (1 is forward, 0 is reverse). We must create an index for each message. We'll simply index our messages counting up from 0. We will also use symbolic constants for each index. We use the #define macro to define the symbolic constants:

```
#define MODULE 2

/* message ids */
#define mSRV1FOR 0
#define mSRV1REV 1
#define mSRV2FOR 2
#define mSRV2REV 3
#define mTIM1    4
#define mTIM2    5
```

The servo messages look like this:

```
/* servo 1 forward */
message[mSRV1FOR] {
  MODULE, cmdSRV_REL, 1, 127, 1
}

/* servo 1 reverse */
```

```
message[mSRV1REV] {
  MODULE, cmdSRV_REL, 1, 127, 0
}

/* servo 2 forward */
message[mSRV2FOR] {
  MODULE, cmdSRV_REL, 2, 127, 1
}

/* servo 2 reverse */
message[mSRV2REV] {
  MODULE, cmdSRV_REL, 2, 127, 0
}
```

Let's look at one of those messages above in a little more detail. As we mentioned earlier, each message block has an index. Here's the first line of the first message:

```
message[mSRV1FOR] {
```

This defines the start of a block with index 0, because we defined mSRV1FOR to be 0. Any reflex vector that contains this index will execute this message when it is activated. Next in the block is this line:

```
  MODULE, cmdSRV_REL, 1, 127, 1
```

The first parameter is the symbolic constant MODULE, which we defined to have the value 2. This is the module ID of the BrainStem. Next in the list is the BrainStem command cmdSRV_REL. This is again a symbolic constant, but unlike the indices we defined, the cmdSRV_REL macro is defined in the file aCMD.tea. The command cmdSRV_REL takes three arguments: 1, 127, and 1. The first value (1) is the servo number, so this command is for servo 1. The next value (127) is the relative position that the servo should rotate to. This parameter is a signed byte, so you are limited to values from –128 to 127. The final parameter (1) is the direction that the servo should rotate. For this command, 1 is forward, 0 is reverse. Note that you must use commas to separate each parameter in the command message. If you leave out any commas, the program will not compile.

Now, we'll work on the timers. We'll have the robot move in each direction for two seconds before reversing direction. We need a two-second timer and, as you'll shortly see, we'll need two of them. The command for the timer is cmdTMR_SET. It has two parameters. The first is the timer number; there are 24 timers that can be used. The second parameter is the number of 0.1 milliseconds to wait. Here are the message blocks that set the timers:

```
/* set timer 1 for 2 seconds */
message[mTIM1] {
  MODULE, cmdTMR_SET, 1, 20000
}

/* set timer 2 for 2 seconds */
message[mTIM2] {
  MODULE, cmdTMR_SET, 2, 20000
}
```

Now we need to construct the reflex vectors. Here is the general form of a reflex vector:

```
vector[index] {
command index, command index, ...
}
```

A reflex vector consists of the keyword vector followed by an index. Within the vector is a comma-separated list of command indices. When the reflex is activated, the commands referenced by the indices will be executed.

In contrast to command indices that can be arbitrary, we sometimes need to use specific indices for reflexes. The BrainStem has a number of predefined reflex indexes. The BrainStem has 24 built-in timers. When a timer counts down, the vector at a specific index is activated. The index 20 happens to correspond to timer 1. (There are 24 timers, so timers 1 through 24 correspond to reflex vectors 20 to 43.) When timer 1 completes, reflex vector 20 is activated. When timer 2 completes, reflex vector 21 is activated.

What we will do is set one of the timers. When that timer counts down, it will cause its associated reflex vector to execute its commands. These commands will move the robot and set a second timer. When that timer counts down, its associated reflex vector will execute its commands to move the robot in the opposite direction, and reset the first timer. However, instead of using 20 or 21 as reflex vector indices, we'll use the constants already defined in aGPReflexes.tea (which we will need to #include into the source file). The constants for the timers are aGP_RFX_TIMER_1 and aGP_RFX_TIMER_2. Finally, each reflex vector will contain the command ID numbers of the commands we want executed. Here is the complete reflex vector for timer 1:

```
vector[aGP_RFX_TIMER_1] {
mSRV1REV, mSRV2FOR, mSRV1REV, mSRV2FOR, mTIM2
}
```

Each vector contains a list of one or more message indexes separated by commas. The message given by the index is executed when the reflex vector is started. In this case, the messages are given by the symbolic constants for the indices 1, 2, and 5. When timer 1 completes, the vector will cause servo 1 to rotate in reverse, servo 2 to rotate forward, and then it will set timer 2. The vector for timer 2 is here:

```
vector[aGP_RFX_TIMER_2] {
mSRV1FOR, mSRV2REV, mSRV1FOR, mSRV2REV, mTIM1
}
```

When timer 2 completes, the vector will cause servo 1 to rotate forward, servo 2 to rotate in reverse, and then it will set timer 1. The reflexes will continue to cycle until you interrupt the program. You may be wondering why each reflex vector repeats the commands for the servos twice. The simple answer is that commanding the servos once doesn't work, and commanding them twice does work.

Here's the final program. The source code for reflexes is stored in LEAF files. We've named this code to_fro.leaf. If you don't want to type this code, you can find it on the CD-ROM in the \Other Applications\ToFro directory.

```
/* file: to_fro.leaf */

#include <aCmd.tea>
#include <aGPReflexes.tea>

/* assume the BrainStem has address 2 */
#define MODULE 2

/* message ids */
```

```
#define mSRV1FOR 0
#define mSRV1REV 1
#define mSRV2FOR 2
#define mSRV2REV 3

#define mTIM1    4
#define mTIM2    5

/************* COMMANDS *************/

/* servo 1 forward */
message[mSRV1FOR] {
  MODULE, cmdSRV_REL, 1, 127, 1
}

/* servo 1 reverse */
message[mSRV1REV] {
  MODULE, cmdSRV_REL, 1, 127, 0
}

/* servo 2 forward */
message[mSRV2FOR] {
  MODULE, cmdSRV_REL, 2, 127, 1
}

/* servo 2 reverse */
message[mSRV2REV] {
  MODULE, cmdSRV_REL, 2, 127, 0
}

/* set timer 1 for 2 seconds */
message[mTIM1] {
  MODULE, cmdTMR_SET, 1, 20000
}

/* set timer 2 for 3 seconds */
message[mTIM2] {
  MODULE, cmdTMR_SET, 2, 20000
}

/************* REFLEX VECTORS *************/
```

```
vector[aGP_RFX_TIMER_1] {
  mSRV1REV, mSRV2FOR, mSRV1REV, mSRV2FOR, mTIM2
}

vector[aGP_RFX_TIMER_2] {
  mSRV1FOR, mSRV2REV, mSRV1FOR, mSRV2REV, mTIM1
}
```

After entering and saving the program above, you would compile it with the leaf command in the Console application.

```
leaf "to_fro.leaf"
```

If the command is successful, a BAG file will be created in the \aUSER directory (for a PC) or as a memo in the MemoPad (for a PDA). The code is downloaded to the BrainStem with the batch command:

```
batch "to_fro.bag"
```

After the code has been downloaded to the BrainStem, we need to kick off the first reflex. This can be done by sending a command from the Console application to the BrainStem. Enter the following in the Console input area, and send it to the BrainStem by pressing Enter:

```
2 38 20 0
```

The first number is the BrainStem module ID. The next number, 38, is the command to start a reflex. The value 20 is the reflex ID. The command can send 1 or 2 bytes of data to the reflex; in this case, we have nothing to send, so the final parameter is 0. Executing the command should start the servos; after about two seconds, the servos will reverse. This will continue until you power off the BrainStem.

Just as with the programs we saw in the first part of the chapter, we can set the BrainStem to automatically start a reflex program. To do that, we need to create a reflex with index 127. Add this reflex vector to the to_fro.leaf program:

```
vector[aGP_RFX_BOOT] {
  mTIM1
}
```

The new reflex vector calls the mTIM1 command message in the program. Recall that this sets the timer for two seconds and then executes the reflex vector aGP_RFX_TIMER_1. After compiling and loading the new to_fro.bag program to the BrainStem, you would use the following two commands to cause the BrainStem to execute reflex aGP_RFX_BOOT the next time the BrainStem is started:

```
2 18 6 1
2 19
```

The parameter 6 is the system variable that controls whether the BrainStem issues reflex aGP_RFX_BOOT at startup. The second parameter instructs the BrainStem to issue the reflex. Instructing the BrainStem to not issue the reflex is done by setting the second parameter to 0, as shown here:

```
2 18 6 0
2 19
```

Programming the BrainStem Using Java

One of the most popular new languages in recent years is the Java programming language and environment from Sun Corporation. Java originated in the mid-1990s as an embedded language for consumer electronic devices. The project was never fully realized and Java became a solution in search of a problem. Coincidentally, the mid-90s was also the time when the World Wide Web began to explode. The features of Java made it an ideal language for programming web applications. While it was first used as a language for programs that ran in a web browser, Java is also ideal for mobile devices (like Palms). Java running on mobile devices is known as Java 2 Micro Edition, or J2ME.

However, there's a slight problem. Java is able to run on many different platforms because the platform-specific code is contained in the virtual machine or in special Java libraries. Host computers communicate with the BrainStem using the host's serial port, and that requires platform-specific code. Sun's Java virtual machine does not come with Palm-specific code for communicating through the serial port. So, even though you can run Java programs on a Palm OS device, you can't use Sun's Java virtual machine for the Palm to control the Palm. If you want to use a Java virtual machine to control the robot, you'll need to run your code on a PC.

Getting the Java Communications API

As we wrote at the beginning of this chapter, we're assuming that you already know how to program. For this section, we're also assuming that you know how to program in Java, and that you already have a Java Software Development Kit (SDK). So we won't be showing you how to get a Java SDK or how to program in Java. We'll simply say that you need one of the Java 2 versions of the SDK (we recommend the 1.4 version) to program for the BrainStem.

However, as we mentioned earlier, you do need to have a special Java library to communicate over the serial port to the BrainStem. This library is the Java Communications API. You can download the Java Communications API at http://java.sun.com/products/javacomm/. At the time we wrote this, Sun provided support only for Windows and Solaris platforms. There is third-party support for Linux; see the FAQ at the Java Communications API web page for a link to the Linux version.

Download and unzip the API to your computer. When you unpack the archive to your computer, it will create a \commapi directory. So we recommend you unpack the archive to your Java SDK directory. For example, if your SDK is installed at C:\jdk1.4, you would unpack the API so that the API files are installed at C:\jdk1.4\commapi.

The installation instructions for the API are contained in a file named PlatformSpecific.html that is included with the installation. We'll briefly repeat the Windows directions here:

1. Copy win32com.dll to the \bin directory of your Java installation.
2. Copy comm.jar to the \lib directory of your Java installation.
3. Copy javax.comm.properties to the \lib directory of your Java installation.
4. Add comm.jar to your classpath.

And here are the directions for Solaris:

1. Add libSolarisSerialParallel.so to the environment LD_LIBRARY_PATH. Alternately, copy libSolarisSerialParallel.so to /usr/lib.
2. Copy comm.jar to the /lib directory of your Java installation.

3. Add comm.jar to your classpath.
4. Copy the file javax.comm.properties to the same directory where you copied comm.jar.
5. Make sure you have the JDK native thread package installed. This implementation only works with native thread. Look at http://java.sun.com/products/jdk/ for details.

You are now ready to write Java programs that can communicate to other devices over the serial port.

Java and the BrainStem

Luckily for us, Acroname has already developed some base classes that handle some of the lower level communication details for us. The class files can be found in the Java software that you can get from Acroname. The files will unpack into the \aJava directory of the BrainStem distribution. The helper classes are located in the \aJava\acroname directory. There is an example program, gp_example.java, in the \aJava directory that shows how to write a Java program for the BrainStem.

Here, we'll show you some excerpts from that sample program so you can get an idea of how you can use the Acroname Java classes to communicate with the BrainStem.

Within your class, there are two primary Acroname classes that you will use: jStem and jGP. The gp_example.java files declares variables for these objects.

```
jStem stem;
jGP gp;
```

Within the constructor, an instance of jStem is created, and then a few lines further, the jGP instance is created:

```
/* build a stem to provide packet protocol */
stem = new jStem();
//code not shown...
gp = new jGP(stem, moduleAddress); //moduleAddress is 2
```

The gp_example.java has a doExample( ) method that actually reads and writes data from and to the serial port. Here are some of the methods used:

```
/* display Analog 0 */
gp.AnalogIn((byte)0)); //returns a float
/* move the servos a bit */
gp.ServoAbs((byte)0, 0.5f); //move servo 0
gp.ServoAbs((byte)0, -0.5f);
```

The AbstractBrainStem class

What we will do first is follow the gp_example class to develop our own base class that can be used by any Java programs to communicate with the BrainStem. We'll write this class to do a couple of things for us:

- Create the jStem and jGP objects
- Provide a method for commanding a servo
- Provide a method for reading a sensor

With these basic tasks provided by the base class, we don't need to re-create that code for every Java class we write for the BrainStem. When we write a Java program for the BrainStem, we extend our base class and our program gets the ability to read the sensors and command the servos.

We'll show the source code first, and then talk about what the class does. To save space, we're not going to show every bit of the class; you can find the complete source code for this class in the \OtherApplications\ToFro directory on the CD-ROM. Here's the base class:

```
public class AbstractBrainStem {
  jStem stem;
  jGP gp;

  public AbstractBrainStem(String portname, byte moduleAddress) {
    InputStream inputStream;
    OutputStream outputStream;

    stem = new jStem();

    try {
      CommPortIdentifier portId =
```

```
          CommPortIdentifier.getPortIdentifier(portname);
        SerialPort serialPort =
          (SerialPort) portId.open("AbstractBrainStem", 2000);

        inputStream = serialPort.getInputStream();
        outputStream = serialPort.getOutputStream();

        serialPort.addEventListener(stem);
        serialPort.notifyOnDataAvailable(true);

        stem.SetStream(inputStream, outputStream);

        gp = new jGP(stem, moduleAddress);
      } catch (Exception e) {
        e.printStackTrace();
      }
    }

    public float readSensor(int sensorNumber) {
      float result = -1;
      try {
        result = gp.AnalogIn((byte) sensorNumber);
      } catch (jErr e) {
        e.printStackTrace();
      }
      return result;
    }

    public void startServo(int servoNumber, float amount) {
      try {
        gp.ServoAbs((byte) servoNumber, amount);
      } catch (jErr e) {
        e.printStackTrace();
      }
    }

    public void stopServo(int servoNumber) {
      try {
        gp.ServoAbs((byte) servoNumber, 0f);
      } catch (jErr e) {
        e.printStackTrace();
      }
    }
}
```

Let's start by looking at the constructor. The constructor here follows the same basic outline as the gp_example class. Your class will need to pass a port name and module address to the constructor. For Windows, the port name will be "COM1" and the default moduleAddress for BrainStem is 2. One of the first things the constructor does is create an instance of the acroname.jStem class. This class is provided by Acroname, and you will find it in the \aJava directory of the Acroname software. Next, the constructor attempts to open the serial port. The CommPortIdentifier and SerialPort classes are part of the Java Communications API. After the port is opened, the code gets the input and output streams for the port and passes them to the jStem instance. After the initialization is complete, the code uses the jStem instance and the module address to create an instance of the jGP class. This class has the methods for communicating with the BrainStem and is used by the other methods of the AbstractBrainStem class.

Next is the readSensor(int) method. This method takes an int that identifies the sensor to read. For the PPRK this argument should have the value 0, 1, or 2. It returns the sensor reading as a float value between 0 and 1.0. The closer the object is to the sensor, the closer the return value is to 1.0. If there is a problem reading the sensor, the method returns –1.

Following the readSensor(int) method is the startServo(int, float) method. This method takes a sensor number as the first argument and commands that servo to rotate. The amount to rotate is given by the second argument, which should be a number between –1.0 and +1.0. Values close to –1 or 1 will cause the servo to rotate faster than values close to 0.

The final method is the stopServo(int) method. This method takes the servo number as an argument and commands that servo to stop.

PPRK Programming with Java

Using our new base class, let's write a Java program for the BrainStem. What we will do is recreate the to_fro.leaf reflex program in Java. Recall that all this program does is command the robot to move in one direction for two seconds, then reverse that movement, and repeat indefinitely. Here's the ToFro.java source code:

```
public class ToFro extends AbstractBrainStem {
  public ToFro(String portname, byte moduleAddress) {
    super(portname, moduleAddress);
```

```
  }

  public static void main(String[] args) {
    ToFro tf = new ToFro("COM1", (byte) 2);
    while (true) {
      tf.moveTo();
      tf.moveFro();
    }
  }

  void moveTo() {
    startServo(1,  1.0f);
    startServo(2, -1.0f);
    sleep(2);
  }

  void moveFro() {
    startServo(1, -1.0f);
    startServo(2,  1.0f);
    sleep(2);
  }

  void sleep(int seconds) {
    try { Thread.sleep(seconds*1000); } catch (Exception ignored) {}
  }
}
```

The ToFro class should be relatively straightforward. The main( ) method calls the constructor for the class, passing the communication port name and BrainStem module ID. The communication port name is, of course, the name on a Windows platform. If you are trying this code on a Mac or Unix platform, you will need to change to the appropriate name for your system. The constructor calls the constructor of AbstractBrainStem, which sets up the serial port to communicate with the BrainStem. The code then starts an endless loop where it calls methods that move the robot in one direction for two seconds, and then in the opposite direction for two seconds.

The next thing to do is to compile our program. To be able to do this, you will need some additional resources. The first thing you'll need is the Java Communications API. We introduced the API earlier in this chapter. If you installed it at that time, you don't need to do anything else. Otherwise, download and install the Java Communications API now. The other resource you need is the BrainStem Java classes from Acroname. Those files can be found in the Acroname Java.zip archive or the Acroname sdk.zip archive. You can get these files from the Acroname web site or on the CD-ROM in the \Other Applications\BrainStem directory.

Set your classpath to include the comm.jar file and the \aJava directory of the BrainStem software. Also, the classpath needs to include the path to the AbstractBrainStem class. Now compiling is as simple as issuing this command:

```
javac ToFro.java
```

After the class is compiled, connect your BrainStem to the computer using a serial cable. You can get a serial cable for this from Acroname. Turn the BrainStem on and execute the ToFro program. If all goes well, you should see the robot start moving back and forth. It will continue to do so until you turn the BrainStem off or the batteries run down.

The Robot Geek Says

BrainStem and Java Tools

All of the software tools for creating, programming, and running TEA programs can be found on the CD-ROM for this book. The TEA software and the tools to compile, load, and launch TEA programs are from Acroname. If you want to search for the latest versions and other information, you should go to Acroname's web site, www.acroname.com.

The tools for writing, compiling, and running Java code can be found at java.sun.com.

We would like to thank the following organizations for making this chapter possible:

- Acroname Corporation, for the BrainStem controller, the TEA language, and SDKs
- Carnegie Mellon, PPRK project, for the original PPRK

Chapter 9

Sensors and Enhancements

Welcome to Chapter 9. If you've gotten this far into the book, it means one of a few things: You are a hardcore roboticist, or you've skipped ahead in the book. You have a working robot, or you're just a speed-reader and your robot is still a collection of parts sitting on the shelf. You're ready to take the next step with your robot, or you're looking for more jokes.

So, let's get the joke out of the way, and we can proceed with the chapter.

Dave: Why do elephants avoid using computers?

Kevin: I don't know. Why do elephants avoid using computers?

Dave: Because they're afraid to use the mouse.

Kevin (and readers): Groan!

Dave: Which just goes to show, there's nothing like a good joke.

Kevin: And that was nothing like a good joke.

Dave: Yes, you're right. My apologies to all the readers out there. Yet, the joke does have some connection with this chapter.

Kevin: I'm interested to hear what that connection might be. But first, how much longer are we going to use this style of writing?

Dave: I was thinking we'd use it for the entire chapter. Then we could turn it into a movie script and become rich and famous or maybe almost famous.

Kevin: And what would be the title of the movie?

Dave: Of Mice and Robots? Sensors and Sensibility? Or maybe, Detector Gadget?

Kevin: I should have known not to ask. Let's just get on with it. What do these awful jokes have to do with this chapter?

Dave: This chapter is all about how to take your Palm Robot to the next level. That is, how to add different input and output devices to your robot. Early in the book, we talked about the basic components of a robot. Here they are again, as a reminder:

- Base
- Processor

- Actuators
- Sensors

There's not too much we need to say about the base. If you've already built the Palm robot, you know all there is to know about the base. (Reminder: Aluminum frame and acrylic deck.) In the last few chapters, we've seen how to program the processor of the robot. That leaves the actuator and sensors—also known as the input and output devices.

Most of the machines we work with have some sort of input and output. That's what makes them useful to us humans. Through a small number of input devices (such as a steering wheel, pedals, or shift lever) a car accepts input and outputs power to move us around. A television accepts input from buttons, dials, and remote controls and outputs video to us. A computer accepts input from keyboard, mouse, and other devices, and outputs data to us.

Likewise, our Palm robot uses input and output devices. The Palm robot uses an infrared ranger to detect distance to objects. It then determines which servos to turn on or off to move the robot in various directions. As we've seen in previous chapters, we can use these rangers and servos to accomplish a good variety of behavior. If you tried some of the sample programs available on the web (see Chapter 5), you've seen the robot follow walls, copy a pattern drawn on its screen, run away from objects, or chase objects—just to name a few behaviors.

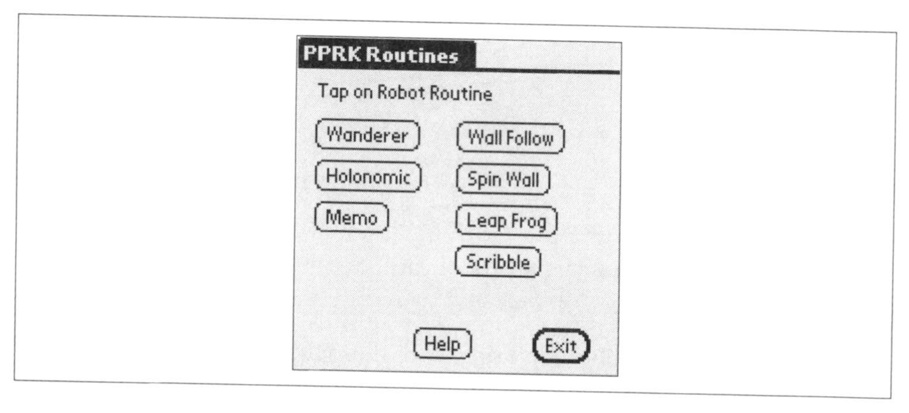

But there are some things our robot cannot do. BEAM robots, which we looked at in Chapter 1, can be built to seek out or avoid light. Our Palm robot can't do that.

Mindstorms robots, built from Legos, can be programmed to follow a line. Our robot can't do that.

Some robots can be programmed to make noise, or even talk. Our robot can't do that.

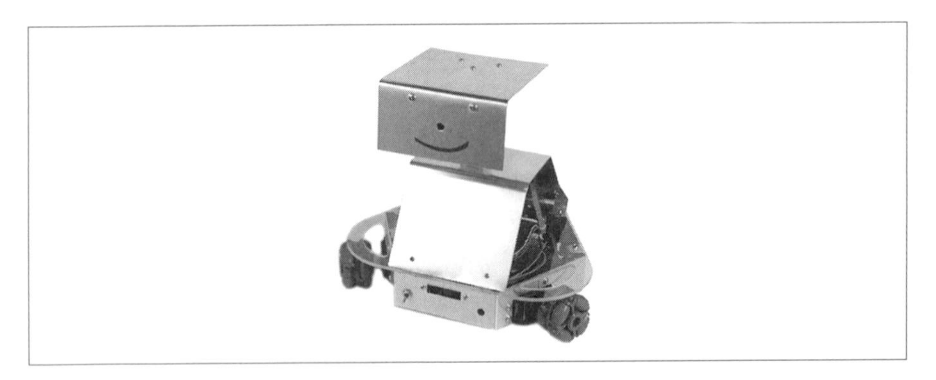

Some robots have death ray lasers that they can use to annihilate enemies (as in Figure 9-1). Our robot can't do that. (And a good thing that is, or life as we know it might end.)

Yes, our robot can't do any of these things. At least, not yet. It is a relatively easy matter to add different sensors and output devices to the controller (either SV203 or BrainStem) of the robot. This will enable you to extend the Palm robot to perform an even greater variety of behaviors than we have already seen.

The SV203 controller has a total of five input ports available for use. If you keep the rangers connected, that leaves two unused input ports that can be

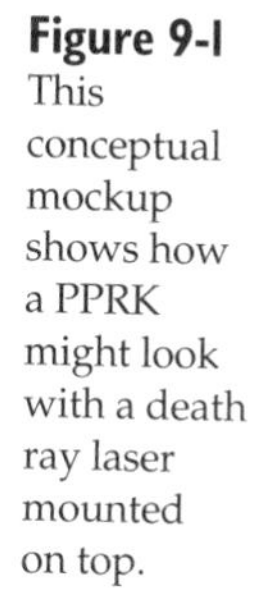

Figure 9-1 This conceptual mockup shows how a PPRK might look with a death ray laser mounted on top.

used by other sensors in addition to the rangers. If you want to use other sensors in place of the rangers, you have up to five available input ports.

The BrainStem controller has five analog input ports, so you can add anywhere from two to five additional sensors. The BrainStem also has five digital ports that can be used for either input or output.

With the SV203, two unused output ports can be used to drive other devices. The BrainStem controller has one unused output analog port, as well as the digital ports mentioned.

Using these ports, we can add all kinds of new capabilities to the robot.

Adding Sensors or Input Devices to the PPRK

When you are learning to program, one of the most important steps you can take is to learn how to get and send data to the user. Your program can loop and iterate and count and add all day long. But until it can get an input from a user and output data to the user, it's not a very useful program. In fact, when we're learning to program a new language, some of the first things we try to learn is how to read input from the keyboard or mouse. Similarly, once you understand the basics of input with the SV203 or BrainStem controller, you can add any number of sensors to your robot.

Adding Sensors to the Pontech SV203

The SV203 has five input ports used for reading data. In this book, we've usually referred to them as analog ports. In fact, they are analog/digital ports—that is, they read an analog value from the input device, and that reading is converted to a 1-byte digital value that is sent to the host computer. Since the value is limited to 1 byte, the value sent by the A/D port is between 0 and 255.

The A/D ports are located on the SV203 in the block of pins labeled J3. The pins are numbered from 1 to 5, starting in the upper-left corner, and moving left to right, top to bottom.

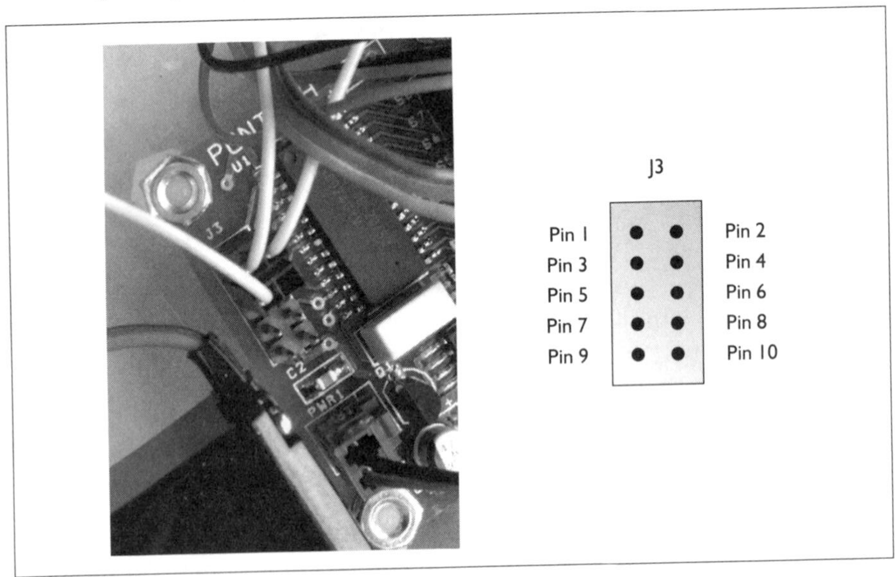

The output line from a sensor device is attached to one of the five A/D ports. Assuming the sensor needs power, the power and ground lines can be attached in a couple of different ways.

With the PPRK we built in this book, the power and ground lines for the rangers were attached to the pins in the output block of ports. Those were the pins labeled S6, S7, and S8 in the block of pins labeled J4 on the SV203. This gave each sensor its own power and ground line.

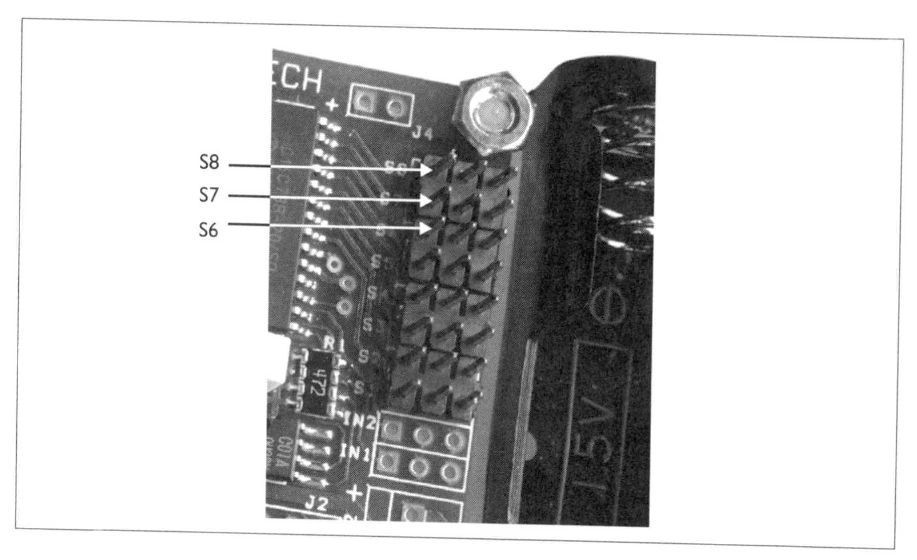

Alternatively, the J3 block on the SV203 also has power and ground pins that you can use for a sensor. These are pins 9 and 10 in block J3. Pin 9 provides a +5V source of power; pin 10 is the ground line.

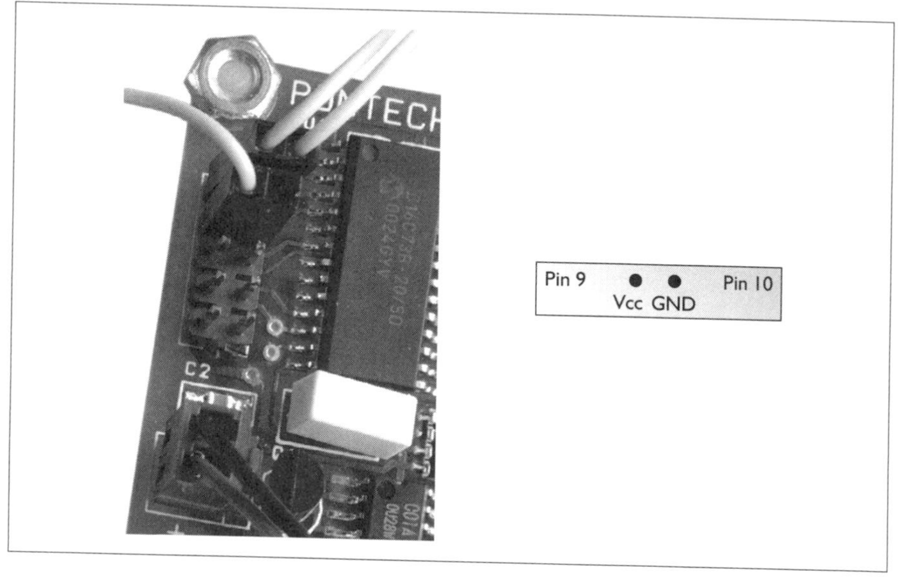

You can conceivably use the power and ground lines in J3 for multiple sensors, but in general, you will find it easier to use the power and ground lines in J4. While it does tie up an output port for each device, using J4 allows you to give each sensor its own source of power. If you try to put multiple devices onto the power and ground lines in block J3, you will need to attach your devices serially or in parallel. When wired serially, the voltage drop across each sensor will not be the full 5 volts supplied by the board. When wired in parallel, each sensor will get the full voltage drop, but current will be reduced, which could prevent the sensor from working properly. So you can see why we prefer to give each sensor its own power line.

Using the power and ground pins from the output ports is exactly what we did when we attached the rangers to the SV203 when building the PPRK (Palm Pilot Robot Kit). Each ranger had a power and ground wire (the black and red wires) that we connected to pins S6, S7, or S8. The yellow wire from the ranger was the output wire. Each yellow wire was attached to one of the pins in block J3.

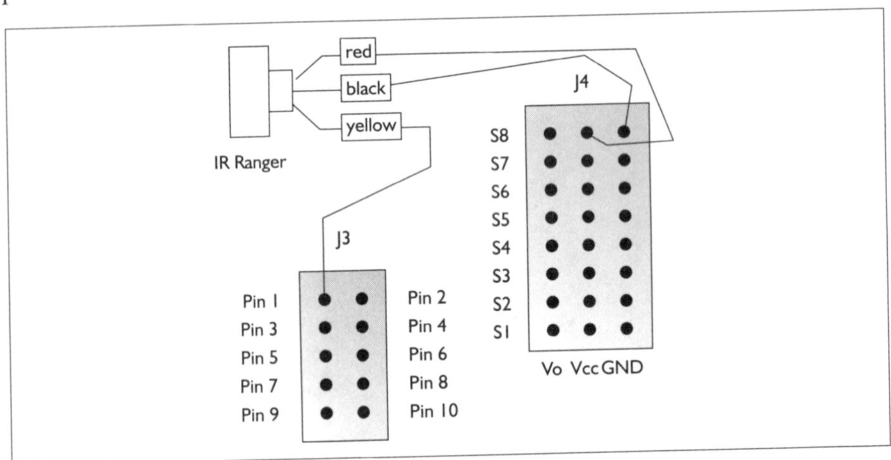

When we attach any other sensor to the SV203, we will follow the same general procedure. Power will be supplied to the sensor using one of the output ports; input from the sensor will be read using one of the A/D ports. Remember, though, if it makes sense for you, you can supply power to your sensor using power from block J3.

Adding Sensors to the BrainStem

With the BrainStem controller, we have a little more flexibility. The SV203 has only analog input ports, the A/D ports in block J3 of the board. The BrainStem has both A/D ports and digital ports. That means, of course, that we can use the BrainStem for analog sensors and digital sensors or devices.

The BrainStem controller contains another port that you can use as well. This is a GP2D02 port located between the digital ports and the servo ports. You can connect another ranger directly to this port. Coincidentally, the name of this IR Ranger is GP2D02. Alternatively, you can attach a special bus called an I2C bus. With the I2C bus, you can connect up to 16 sonar rangers that can also detect light, a compass module, and a text-to-speech module (SP03), all on the same bus. Discussing the I2C bus further is beyond the scope of this book; however, once you are familiar with the BrainStem, you may want to investigate using the GP2D02 port and an I2C bus to attach other devices to the robot.

You should already know where the A/D ports of the BrainStem are located (see Figure 9-2). These are the five sets of pins, of which we used three for the rangers.

These five sets of pins are located on the upper-left edge of the BrainStem controller board. They work similarly to the A/D ports on the SV203. With the BrainStem, the ports can read an analog voltage from 0 to 5 volts on any of the ports. Like the SV203, this is converted to a digital value. However, in the case of the BrainStem, the ports can output a 10-bit number, which means the output value can range from 0 to 1023. So when you read the A/D port with TEA or Java, the returned value will be between 0 and 1023.

Figure 9-2

The input and output ports of the BrainStem are on the top edge of the BrainStem circuit board. In the orientation shown, the analog input ports are on the right, the servo ports are on the left, and the digital ports are in the center.

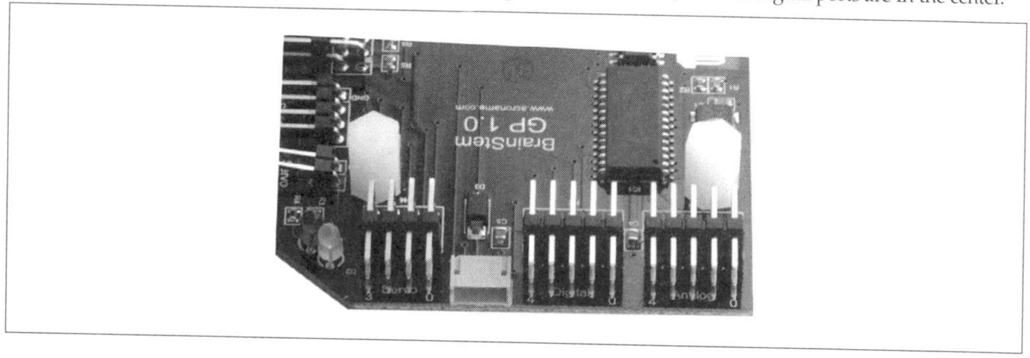

Another difference between the BrainStem and the SV203 is that the pins that provide the power and ground lines to the sensor are located next to the input pins. Each input pin gets its own power and ground line, so the problem of sharing the power and ground lines, or stealing power and ground from the output ports that we had with the SV203, is not an issue with the BrainStem.

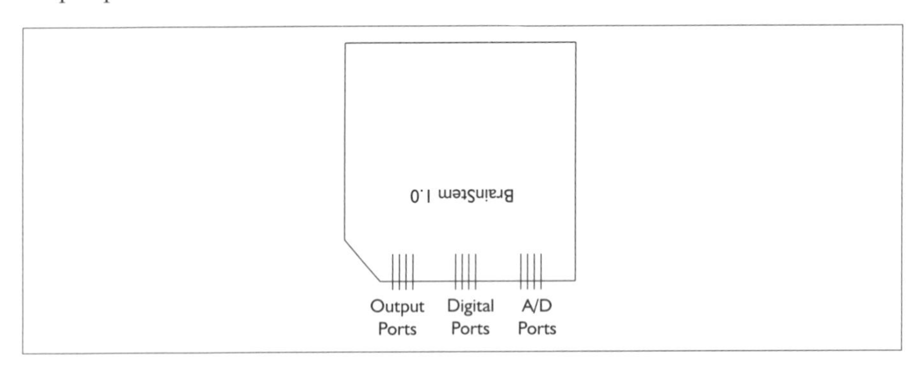

Finally, the BrainStem also has five digital ports that can be used as either input or output ports for digital devices (see Figure 9-2). These ports are configured to be digital input ports as the default. You can change them to output ports through software. When configured for output, the on state of the port will be using about 25 mA (milli amperes) of current. These outputs can be used to drive any digital device that can accept 25 mA of current. When configured as input ports, they can measure the on or off state transmitted by a digital component. You can attach any device that outputs a digital signal that can be read on the digital port. For example, you could hook a radio control receiver to the digital input ports and use the receiver to inject input into the robot for driving the robot or testing its operation. In addition, ports 1 through 4 can be configured as high-resolution timers that can be used to measure pulse widths or elapsed times.

So, like the SV203, we can add two to five analog sensors to the BrainStem, and one to five digital sensors. That makes ten total sensors, and none of them steal one of the output ports for its power.

Using a Line Detecting Sensor

Researchers who study robotics have found robotic movement to be one of the more difficult problems. While walking seems extremely simple to most of us, re-creating in a robot the ability to sense and move through its environment is extremely complicated.

Walking robots have been around for years, but most of them have more than two legs. It's just much simpler to maintain balance when your robot has four or more legs. It's only recently, with Honda's Asimo as one of the best examples, that robots have been created with autonomous two-legged motion.

Robots with wheels don't have to worry about balance and walking, but like walking robots, they have the same challenge of needing to sense their environment and maneuver through that environment. To solve this problem, various approaches have been attempted by researchers. One approach is to limit the boundaries of the problem. By limiting the scope of the problem, the solution becomes simpler.

What is that limited scope? From the title of this section, you might guess that it involves following a line. And you would be correct. By limiting the challenge from free maneuver through the entire environment to movement along a predefined track or line, our robot becomes much simpler to design. Creating a robot that successfully navigates a line is much easier to build than a robot that has the ability to meet all the challenges of its environment.

To be sure, making that limitation is not always possible. The Mars Sojourner robot, a robot that needed to operate semiautonomously in its environment—the surface of Mars—would not have been viable if all it could do was follow a line. But since our robot is not going to Mars, we can use this solution.

We tried out two different sensors with our robot. The first is a reflective infrared (IR) sensor made by Lynxmotion, Inc. (www.lynxmotion.com). The second is a solid-state photoreflector from Hamamatsu Photonics (www.hamamatsu.com). Both sensors are available from Acroname.

The Hamamatsu Photoreflector

The photoreflector from Hamamatsu is similar to the infrared rangers that we've already used. It has an IR light emitting diode (LED) that emits an infrared signal. When this signal is reflected by some object, another component detects the signal. Unlike the ranger, however, this sensor signals only whether or not it receives the IR pulse, but it does not calculate anything else from the received signal.

This sensor comes in two varieties. If we assume that we have a black line on a white surface, one sensor sets the output to a high state (close to 5 volts) when it detects a reflected IR signal; otherwise, it sends out a low. That is, when the sensor is over a white or reflective surface (not over the line), its output state will be 5 volts. This sensor has the model number P5588. The other sensor is just the opposite. It sets the output to a high state when it does not detect a reflective IR signal; otherwise, it sends out a low. That is, when the sensor is over the line, a black surface, its output state is 5 volts. This is the P5587 photoreflector sensor. These values are illustrated in the following table.

Sensor	**Output**	
	Over Black	**Over White**
P5587	5V	0V
P5588	0V	5V

This sensor is easily connected to either the SV203 or the BrainStem controller. The sensor has five pins, but we care about only three of those pins. As with the IR rangers, we need the power input (also known as Vcc), the ground pin (GND), and the output pin (Vo).

When attaching the Hamamatsu photoreflector to the SV203, the Vcc and GND pins are attached to the Vcc and GND pins in block J4. The Vo pin is attached to one of the A/D ports in block J3. This is the same way we attached the IR rangers to the SV203. We recommend that you use a circuit board or breadboard to mount the sensor. This will give you a way to attach the sensor to the frame of the robot. For best operation, the sensor needs to be mounted within just a few millimeters of the surface. The specification sheet that comes with the sensor gives the operating characteristics at 3mm. That is approximately 1/8 inch. The sensor will work if it is slightly further away than the optimal 3mm, but its performance will be degraded.

Recall from Chapter 7 that we can read the analog port with the command AD*n*, where *n* is the port number from 1 to 5. Thus, to use this sensor, you will send the AD command to the board and read the output. If the sensor is in the high state, you will receive a value of 255. When the sensor is in the low state, you will receive a very low value (but not necessarily 0). A representative code listing to do this using HotPaw Basic is shown next. This code listing assumes that the P5587 IR photoreflector is connected to port 1. You would need to change the listing to reflect the actual port you use.

```
print #5, "BD1 AD1" : rem tell board sensor 1
result = get$(#5, 0) : rem read 1 byte
if (result < 50)
  /* P5587 sensor is not over line */
else
  /* P5587 sensor is over line */
endif
```

Connecting this sensor to the BrainStem is just as easy. Just as with the SV203, you will connect the GND, Vcc, and Vo pins of the sensor to the correct pins of the BrainStem.

After connecting the sensor to the BrainStem and mounting it to the robot at approximately 3mm over the surface to be sensed, we can read its output using the TEA language. Here is sample TEA code for the BrainStem that assumes the IR photoreflector is connected to port 1.

```
r1=aA2D_ReadInt(APPRK_IR1);
if (r1<50) {
```

```
  /* P5587 sensor is not over line */
} else {
  /* P5587 sensor is over line */
}
```

The Lynxmotion Single Line Detector

The Single Line Detector from Lynxmotion is similar to the Hamamatsu photoreflector—basically, the sensor includes an IR emitter. When the sensor is positioned over a reflective surface (such as a white surface), the IR signal is reflected back to the sensor. When the sensor detects a reflected IR signal, its output is high, or close to 5 volts. When the sensor is positioned over nothing, or over a surface that does not reflect the IR signal (such as the black line), the output of the sensor is low, or close to 0 volts.

We recommend that you purchase this detector from Acroname. The detector is already set up to connect to your BrainStem controller. The Vcc, Vo, and GND lines from the sensor are already inserted into a Molex housing, which can be directly attached to the A/D port of the BrainStem controller.

After attaching the Lynxmotion Single Line Detector, you would use the same TEA code used previously to read the sensor. If the sensor returns a high value, the sensor is not positioned over the line; if the sensor returns a low value, the sensor is positioned over the line.

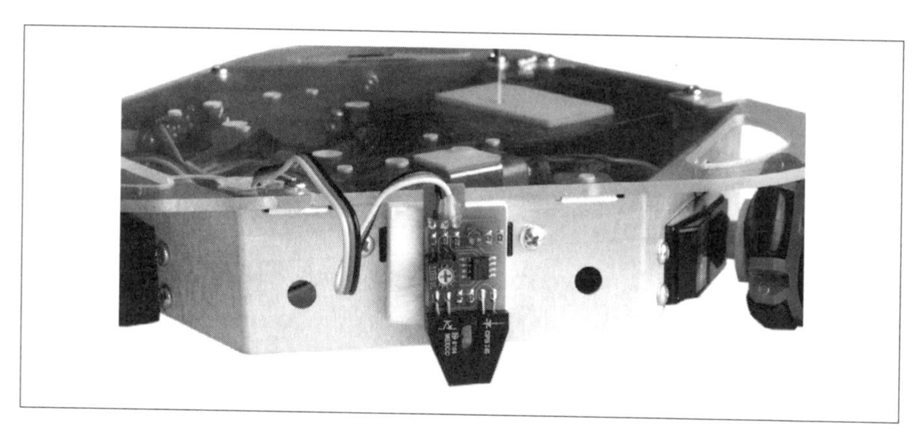

Using a Temperature Sensor

Another sensor that we tried with the PPRK is the LM34DZ temperature sensor. This sensor is made by national Semiconductor. A simple search of the web should turn up several suppliers. We found some of these at www.allelectronics.com and www.pigselectronics.com.

As with the photo detectors, the LM34DZ can be directly connected to either the SV203 or the BrainStem. The LM34DZ has three leads: one for Vcc, one for Vo, and one for the GND. These are connected directly to the appropriate leads on either the SV203 or the BrainStem.

After the sensor is connected to the SV203 or the BrainStem, you can read it using the same techniques we've already discussed. The LM34DZ changes its output voltage based on the temperature it senses. For each degree Fahrenheit that is sensed, the sensor outputs 10 millivolts. For example, if the temperature is 65 degrees, the sensor will output .65 volts; a temperature of 75 degrees will cause the sensor to output .75 volts. The A/D ports of the controller will convert the voltage to a digital value between 0 and 255.

Other Sensors

At this point, attaching other sensors to either the SV203 or the BrainStem controllers should be a simple matter for you. If the sensor already has pins for Vcc of 5 volts, a GND, and an output voltage line Vo, you can attach it directly to either controller using the same techniques we've used in this chapter. If you need more information, one of the best resources we've found for modifying the PPRK with additional sensors is Roberts Gadgets and Gizmos at www.bpesolutions.com/gadgets.ws/gproject3.html.

Adding Output Devices to the PPRK

We've finished the first step. We know how to attach sensors to the PPRK and read the data from those sensors. Now we need to look at attaching output devices other than servos to the controller and using those other devices.

Adding Output Devices to the Pontech SV203

The SV203 has eight output ports that can be used. The output ports are located on the SV203 in the block of pins labeled J4. The pins are numbered from S1 to S8, starting at the bottom of the block of pins.

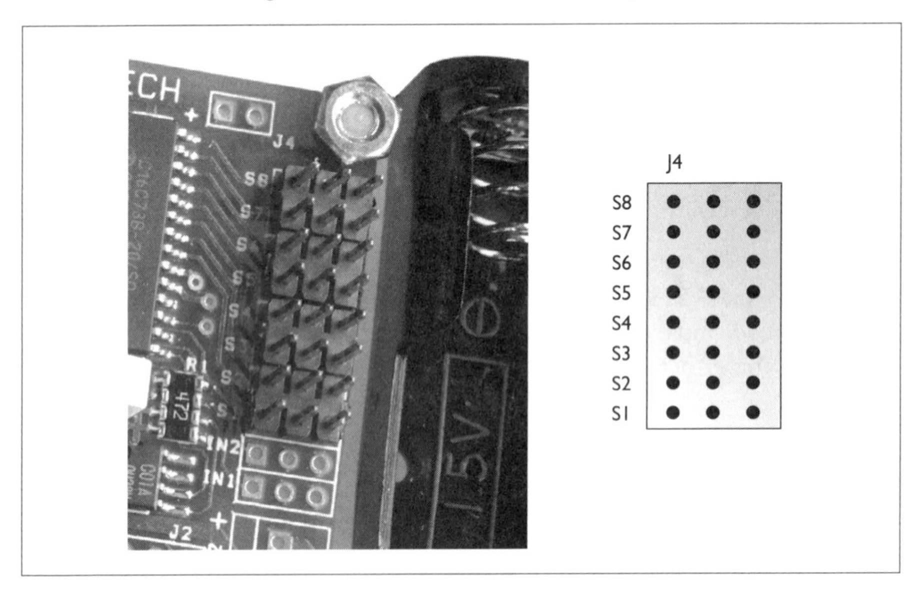

Unfortunately, three of the output ports are used by the IR rangers of the PPRK. With the PPRK we built in this book, the power and ground lines for the rangers were attached to the pins in the output block of ports. Those were the pins labeled S6, S7, and S8 in the block of pins labeled J4 on the SV203. Thus, assuming that you are going to keep the servos attached to S1, S2, and S3, only two output ports are available for attaching additional devices.

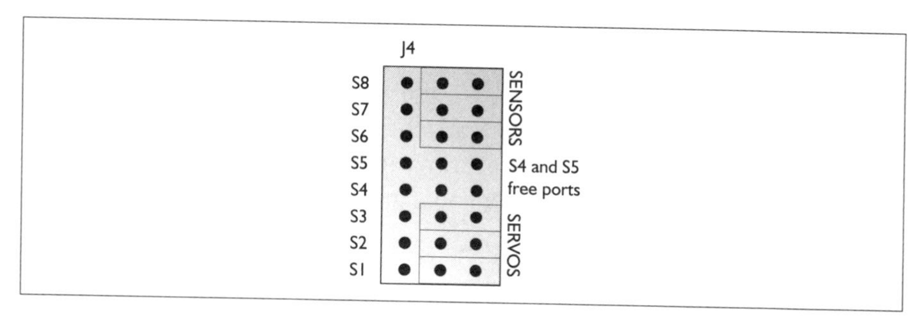

When we attach any other output device to the SV203, we will follow the same general procedure we used for the servos. The GND of the device is attached to the GND of the SV203 (the pin closest to the edge of the board). The power line of the device is attached to the middle pin. The signal line of the device is attached to the pin to the inside of the board.

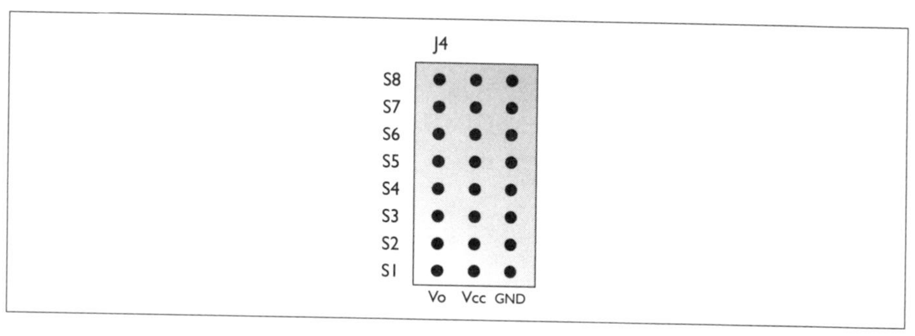

Adding Sensors to the BrainStem

With the BrainStem controller, we have a little more flexibility. The BrainStem SV203 has only four servo ports, but it has an additional five digital ports that can be used for output. That means that even if we keep the three servos, we can attach up to six other output devices to the BrainStem.

You should already know the location of the output ports of the BrainStem (see Figure 9-2). These are the four sets of pins on the right side of the upper edge of the BrainStem. We used three of these sets for the servos. As with the SV203, one pin is used for the GND, one supplies power to the device, and the third is the signal line. With the servos, the signal line passes commands to the servo to cause the servo to rotate.

LEDs

The first output project we will look at is an extremely simple one. Light Emitting Diodes, or LEDs, can be directly attached to either the SV203 or the BrainStem. (Now these are not exactly the same as death ray lasers, but if you squint your eyes a little, you can pretend they are almost the same.)

We started by connecting an LED to the analog output ports of the SV203 and the BrainStem controller. Later in this chapter, in the section "Using Digital Components with the BrainStem," we'll connect the LEDs to the digital output ports of the BrainStem.

We used simple colored LEDs available from various sources. The ones we used are from Radio Shack, part numbers 276-022 and 276-041.

LEDs have two leads—one known as the *anode* and one known as the *cathode*. If you look at the LEDs, you'll see that one lead is shorter than the other. The shorter lead is the cathode and the longer lead is the anode. The anode is attached to the signal, or Vo, line. On both the SV203 and the BrainStem, this is the pin farthest from the edge of the board. The cathode is attached to the GND line of the controller. For both the SV203 and BrainStem, this is the pin closest to the edge of the board.

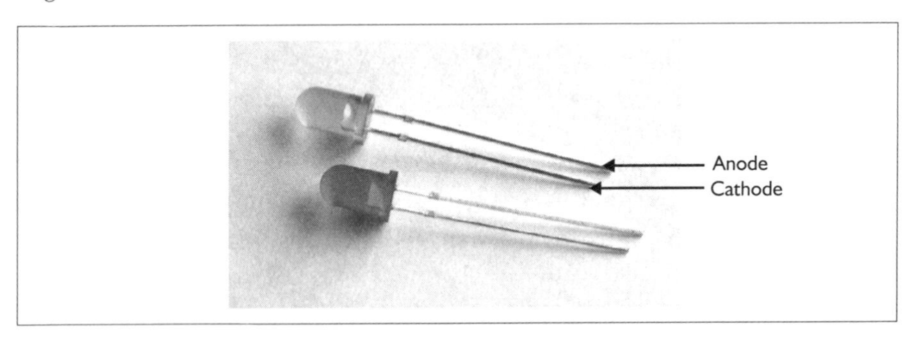

Using the programming we saw in Chapter 7 and Chapter 8, we can drive the LEDs so that they light up on command. With the SV203, we will use the SVn Mn command. Recall that the SVn command is used to select an output port; the Mn command sets the output port to a particular value from 0 to 255. The value 255 will cause the LED to light. For example, this line of BASIC code will cause an LED connected to output port 4 to light:

```
print #5, "SV4M255"
```

Similarly, if we've attached an LED to output port 4 of the BrainStem, this line of TEA code will cause the LED to light:

```
aServo_SetAbsolute(3, 255)
```

Making Noise with Sound Devices

If you have a BrainStem, you might also want to check out the Devantech SP03 Speech Synthesizer (www.robot-electronics.co.uk/). You can get this device from Devantech or from Acroname. While this device is not cheap, who doesn't want a robot that can talk? The synthesizer has a built-in speaker. All you have to do is add digitized phrases, and command the synthesizer using your Palm or PC.

The SP03 can store up to 30 phrases consisting of up to 1925 characters. Alternatively, you can send text to the SP03 and it can convert the text to speech. Full details and software for using the SP03 can be found at both Acroname's and Devantech's web sites.

Other Output Devices

At this point, attaching other output devices to either the SV203 or the BrainStem controllers should be a simple matter for you. If the device needs power, you will connect its power input to the Vcc line of the controller. The signal line is connected to Vo and the ground line to GND. If your output device does not need power, as with the LEDs, you will just connect the Vo and GND lines. If your device does not conform to these specifications or it uses different voltages, you can find more information on how to connect devices to the SV203 or BrainStem at Roberts Gadgets and Gizmos at www.bpesolutions.com/gadgets.ws/gproject3.html.

Using Digital Components with the BrainStem

Any device that can accept digital signals can be used with digital ports of the BrainStem controller. For example, the LEDs that we connected to the analog output ports earlier can be connected to the digital ports of the BrainStem (see Figure 9-2).

As with the output ports, you will connect the anode (the longer lead) of the LED to the digital signal line of the digital port. This is the pin farthest from the edge of the board. The cathode will be connected to the GND pin, the pin closest to the edge of the board.

You can then set the port to high and light the LED with these TEA commands:

```
aDig_Config(0, ADIG_OUTPUT)
aDig_Write(0, 1)
```

Where to Go From Here

As far as this chapter goes, we've barely touched the surface of what you can do with your Palm robot. The number of devices you can attach to the robot is limited only by your imagination.

We've looked at input devices that are analogous to the sense of sight. But we humans have at least four other senses. Sensors that detect the presence of particular molecules in the air are out of our reach, but sensors that mimic touch or sound could easily be added to the robot. But don't let all this stop you. Just because humans are limited to five senses doesn't mean your robot is limited to those same five senses.

We've also spent some time investigating how to add other output devices to the robot. Our little PPRK is limited in what it can do—it's not going to pilot a Mars rocket or control the power grid of New England. On the other hand, it makes a great platform for investigating simple control and output applications. You can modify and program your robot to do many other things as well.

As far as the Palm robot, our journey is nearly at an end. We've spent a lot of time looking at how to build, program, and extend the robot. But we're more than robot fans (that is, enthusiasts—not rotating blades that provide cooling air), we're also Palm enthusiasts. Working endlessly day and night on just the PPRK is not our only idea of fun. Palm OS devices can do more than control your PPRK, and in the next chapter, we'll take a quick look at some of the other robotics applications that you can use your Palm to accomplish.

Chapter 10

Having Fun with Your Robot

Your Palm robot may be alive and well, but how can you claim to live the robot lifestyle without surrounding yourself with robot stuff? It's like Kevin always says: "I need Powerpuff shower curtains to go with my Powerpuff toothbrush and Powerpuff bathrobe." You can apply his dedication to that particular hobby to the world of robots.

Where should you start? Right on your PDA, of course. In this, our final chapter, we'll point out some of the coolest robot stuff you might want to get to impress your family and friends. And it'll give you something to do when you're not playing with your Palm robot.

Installing Palm Programs

You can find all of the Palm OS programs mentioned in this chapter at www.palmgear.com, an excellent resource for shareware, freeware, and commercial PDA software. To track down one of the titles we discuss, visit Palmgear and enter the name of the program in the search box. When you get to the proper page, be sure to download the appropriate file—usually .zip for Windows users or .sit for Mac users.

If you're new to PDAs, it pays to keep in mind that downloaded applications aren't immediately ready for installation. Usually, programs arrive on your PC in the aforementioned form of .zip or .sit files. You will need to expand the .zip or .sit file into its original form before the program can be installed on your Palm (see Figure 10-1).

We recommend these tools for managing compressed files:

- **Windows** Use WinZip to uncompress ZIP files. If you're using Windows XP, a rudimentary unzipper is built in, so you don't even need WinZip.
- **Macintosh** Use Aladdin StuffIt Expander to manage SIT files.

Of course, if you're already happily expanding compressed files with another program, keep up the good work; these are just our favorites. If you need to find WinZip or StuffIt Expander, try www.download.com.

Once expanded, a typical ZIP file will have many individual files inside—readmes, HTML files, and so forth. You can ignore most of that stuff. The readme will have instructions you may need to read, of course, but in general you need to install only files that bear the file extensions .prc (which are the executable program files) or .pdb (which are database files that supply information the PRCs need to run).

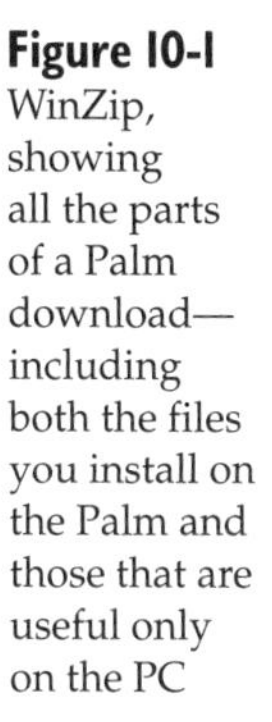

Figure 10-1 WinZip, showing all the parts of a Palm download—including both the files you install on the Palm and those that are useful only on the PC

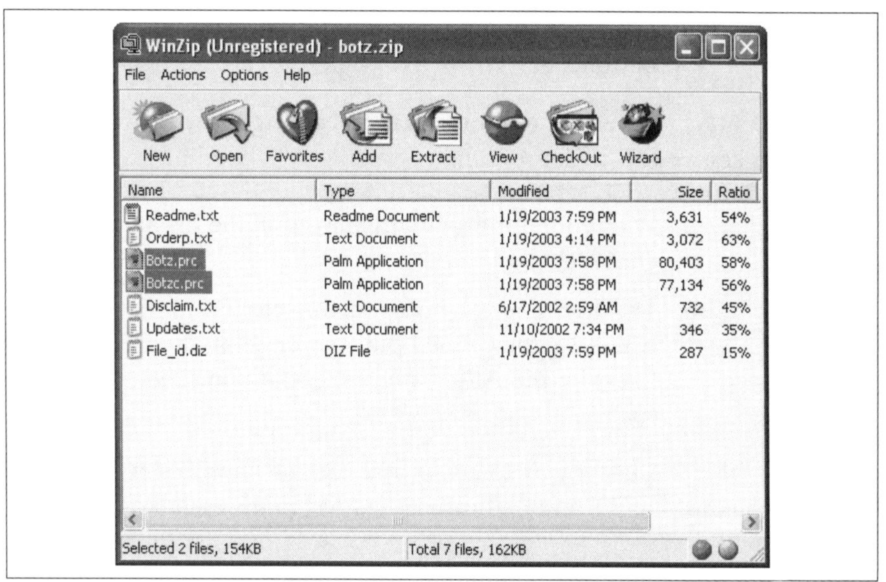

Once you've got a PRC ready to install, do this:

1. Start the Palm Desktop by choosing Start | Programs | Palm Desktop | Palm Desktop.
2. Click the Install button on the left side of the Palm Desktop screen. The Install tool appears.
3. Click the Add button. You see the Open dialog box for selecting applications.

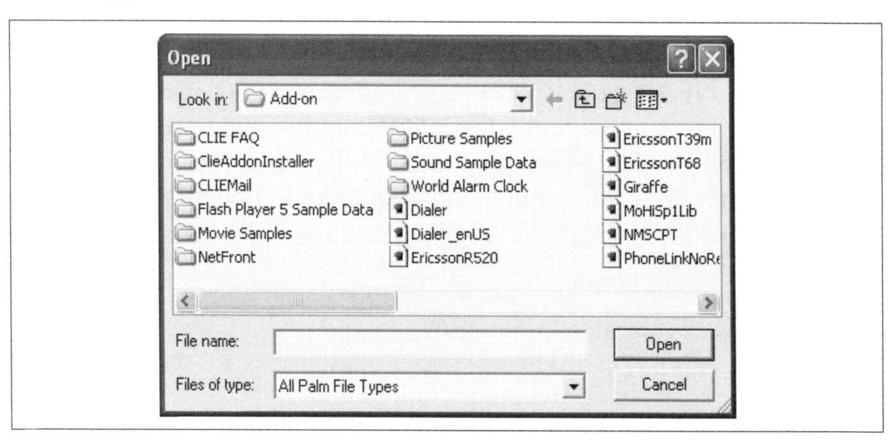

4. Locate the program you want to install and select it. Click the Open button.

TIP: You can select multiple applications at once by holding down the CTRL key as you click programs in the file list.

5. With your application displayed in the Install Tool dialog box, click Done.
6. The next time you hot sync, the selected application is installed, usually to the Unfiled category on your Palm. If you don't see it there, look in All—some apps get automatically filed elsewhere, such as in Games or Main.

TIP: A faster way to install apps on the Palm is simply to double-click the icons. They'll automatically be added to the install queue.

Mindstorms from Your Palm

Lego Mindstorms have proven incredibly popular with kids, adults, roboticists, educators, and just about everyone in between. It was only inevitable, then, that eventually someone would create a Palm-based tool for interacting with Mindstorms.

Unfortunately, the PBrick project, developed by Dr. Ing. Till Harbaum, isn't intended for nonprogramming-savvy users. In a nutshell, Harbaum has written a library that provides an application programming interface (API) for two-way infrared communication between the Palm PDA and the Lego Mindstorms RCX controller.

What good is it? If you're a programmer, you can take advantage of this library to emulate the Lego Mindstorms remote control, program the RCX unit, or even use the Palm as a CPU for advanced on-board logic and control. Using the Palm as the robot's CPU could potentially give your robots a lot more flexibility and intelligence than what's possible with the standard Lego RCX.

If you're not a programmer but you want to experiment a bit with the PBrick Library anyway, you can: the software comes with a pair of demo programs that give you a taste of what's possible with the software. First, download the software from www.harbaum.org/till/palm/pbrick.

You need to install three files on your Palm: the PBrickLib.prc file itself, as well as PBRemote.prc and PBDemo.prc. Here's what they do:

- **PBRemote.prc** This program emulates the Lego Mindstorms remote control, and it can be used to operate the RCX unit. You can use the remote to operate the motors, for instance, as well as activate whatever programs are stored in the five storage blocks.
- **PBDemo.prc** This demo program actually uses two-way communication with the RCX unit and shows a lot of the power the software is capable of. The FW button queries the RCX unit to determine the firmware version number, for instance, and the BT button determines the battery level.

NOTE: When we tested the PBrick library, it wasn't compatible with Palm OS 5 devices like the Tungsten T. We ran it successfully on a Palm m515, though, so it should run fine on almost all older devices.

What Is Lego Mindstorms?

If you haven't experienced Lego Mindstorms, you're missing out on something very cool. Yes, they're Legos—but they're not just for kids. The Mindstorms Robotic Invention System 2.0 is an open-ended set of robotics tools that lets you create a wide-ranging variety of robots and program them to interact with their environment and other Mindstorms robots. The heart of every Mindstorms robot is the RCX (which you can see in the next illustration in the image of a typical Mindstorms robot), which houses the battery pack, motor and sensor ports, infrared transceiver, and a CPU. After you build your robot, it can be stationary, or it can walk or roll—you program it using a visual, drag-and-drop environment on the

PC. Programs are beamed via infrared to the robot, which then can run the program. The core system comes with touch and light sensors, but other sensors, such as rotation and temperature sensors, are also available.

FIRST Scoring

If you are into robots, you no doubt know all about FIRST (For Inspiration and Recognition of Science and Technology). FIRST sponsors an annual robotics competition that currently features 20,000 students in 800 teams across the country and internationally. FIRST also sponsors a Lego league that allows students to compete with Lego Mindstorms designs—it's considered a sort of "little league" in the FIRST community.

If you're involved in the FIRST Robotics competition at any level, you'll want to check out Scoring 2.00, a free application that's a scoring calculator for the 2002 event. If you're a competitor, you can use the calculator to help determine smart competition strategies.

Play a Few Games

All work and no play makes for a boring roboticist. We've rounded up about a dozen robot games for you to try on your Palm OS PDA. We tested these games on a shiny new Tungsten T, so most of them should work with pretty much any Palm model—except perhaps for the very oldest devices that predate Palm OS 3. We'll point out exceptions when we run into them.

Flying Robots

Okay, to tell you the truth, we're not really sure why it's called Flying Robots. Honestly, it looks a lot more like a variation on the old Space Invaders arcade game. The idea behind the game, though, is that a slew of flying robots escaped from the factory at which they belong (really, ripped from today's headlines, when you think about it), and it's your job to return them. It looks, feels, and smells a lot like Space Invaders, so just use the buttons on the Palm or the stylus to move your character around the screen and fire.

Actually, it's a pretty neat game because it's written in Java. That means you'll need a Java interpreter to play the game, and anytime you have to install arcane programming tools on your Palm just to play a game it is a good day indeed. In addition to the game itself, you need to visit this URL and download MIDP for Palm OS: http://java.sun.com/products/midp4palm/download.html.

After you download the ZIP file onto your PC, you can extract MIDP.PRC and install it on your Palm. Throw away the rest of the ZIP file—you don't need it. Then start blasting robots.

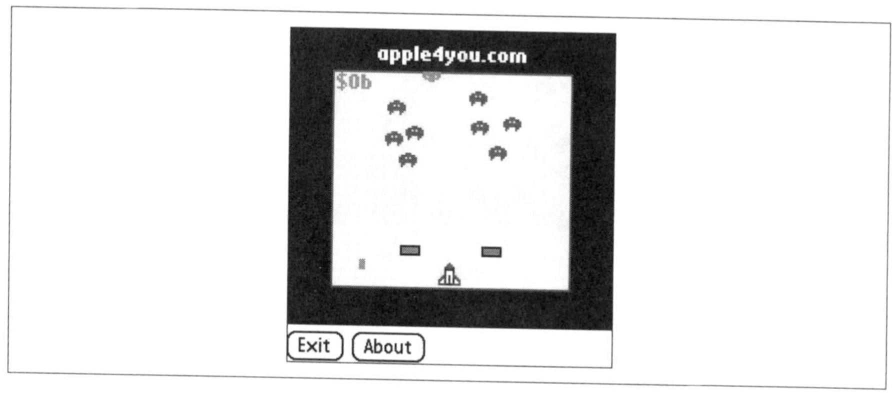

Welded Warriors

If, like us, you're convinced that robots should always be equipped with some sort of death ray, you might appreciate Welded Warriors, a cool robot combat game that more or less follows in the tradition of Mech games. In a nutshell, you control a giant robot that fights battles in a large combat arena for money and fame.

This ambitious game is part fighting, part role playing. You can visit locales like the junkyard and machine shop, where you can purchase parts such as power supplies, shields, and lasers for your robot. Then you equip your robot with these devices and challenge someone to a fight (in the bar, of course).

Actual combat is turn-based; you tap offensive actions such as kicking, punching, or firing your laser to inflict damage, and you defend yourself to conserve power and recharge your shields. The full version of the game—which costs about $15—features two dozen robots and 20 worlds to visit. It's an entertaining way to pass the time before your first morning meeting.

Botz

In the annals of computer gaming, two crazed-robot titles leap to the forefront: Daleks and Berzerk. Both are mindlessly simple games from the earliest days of arcade gaming. Berzerk pitted you against a horde of zombie-speed robots that meandered toward you as if their legs were filled with molasses, while you tried valiantly to work your way through a maze-like set of rooms.

Even though no one moved very fast, it was a gripping arcade experience anyway, probably because the robots taunted you with the ominous threat that they would, indeed, get the humanoid (that audio taunt was arguably the game's coolest feature). And eventually, they always did.

The other great robot classic was Daleks, a Doctor Who-derived title for the Macintosh in which you wandered around a field full of Delek robots, and they inevitably converged on you with a single-minded obsession that caused them to bump into each other in the process, annihilating themselves. Therein was the solution: Daleks was something of a puzzle game, in which you moved around the screen to get the robots to run into each other. Your sole weapon—a teleporting Sonic Screwdriver—was always a last resort.

All that's a long-winded way of getting to the fact that Botz combines both gaming experiences into a single treat for the Palm. In real-time mode, Botz simulates a game of Berzerk, in which you run away from the persistent-but-sluggish robots bent on your destruction. In turn-based mode, you try to collect goodies from the screen while helping the Dalek-like robots run into each other. The full version of Botz is $9.95.

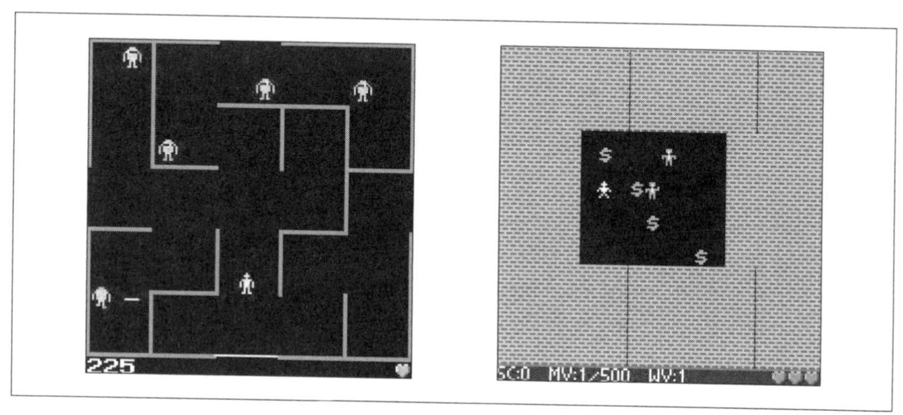

Nanobots

In many ways, robotic hardware is the easy part of designing and building a robot. Programming a robot to respond intelligently to its environment without being programmed with contingencies on how to deal with every single aspect of the world is the hard part. And many roboticists, of course, realize that traditional programming techniques simply won't cut it.

That's where *neural networks* come in. When neural nets burst onto the scene a few decades ago, some scientists thought that they would give machines human-like intelligence in short order. Since neural nets are learning programs that improve through training and experience, they bear a passing resemblance to the way the human brain works. And technologies saw neural nets as the key to unlocking voice translation, pattern recognition, auto-navigation, and a slew of other problems.

It really didn't pan out that way, however; neural nets, despite their early promise, haven't proven to be the magic bullet that solves all of our deepest programming challenges. Nonetheless, they still have a role in robotics, and many roboticists continue to develop robots using neural networks. If you want to get a taste of what neural networks are all about, try Nanobots, a clever strategy game that's also a primer on neural networks. You need to teach a batch of nanobots how to eliminate a virus by moving around the screen. There are other considerations: you need to minimize your interaction with healthy blood cells and other nanobots. The program costs a mere $4.95.

To use Nanobots, you'll need to install a small utility called Mathlib on your Palm; it's free and you can find a copy of that program on Palmgear as well.

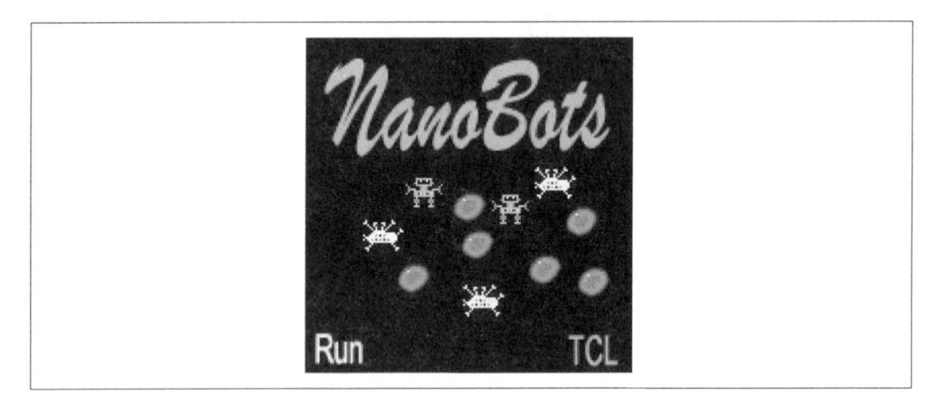

YADaleks+

This version of Daleks is even better than the mini-Daleks game you'll find in Botz, and it's well worth checking out.

To be perfectly honest, versions of Daleks are only a little less common than fireflies in June. We found a small assortment of Dalek-like games, but trust us: this one, with an abbreviated title that stands for "Yet Another Daleks," is the one to try.

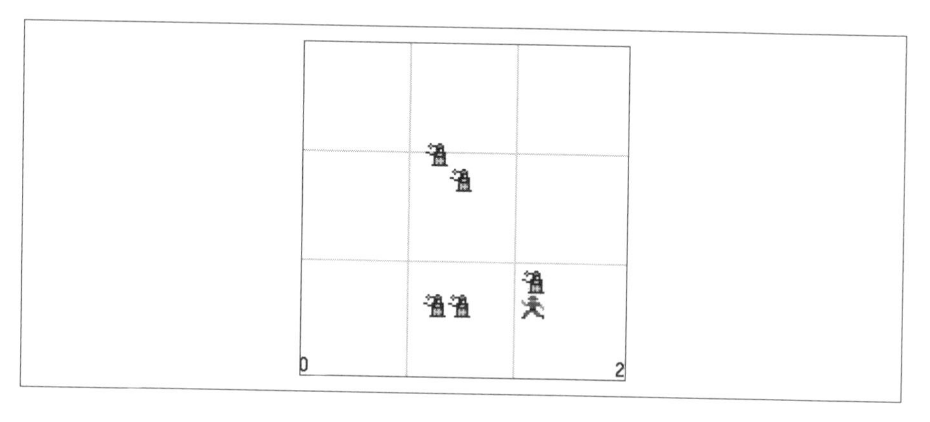

Minebot

These days, it seems like most "robots" are little more than glorified radio-controlled cars, sometimes equipped with a lot of armor, sensors, and manipulating arms. Robots like these are used by scientists, the military, police, and emergency workers to go places and do things too dangerous for humans.

Minebot puts you in the driver's seat of just such a robot: you're tasked with traversing a minefield to rescue hostages. You are equipped with a proximity scanner to sense the presence of nearby mines, and some levels include the use of dynamite and other tools.

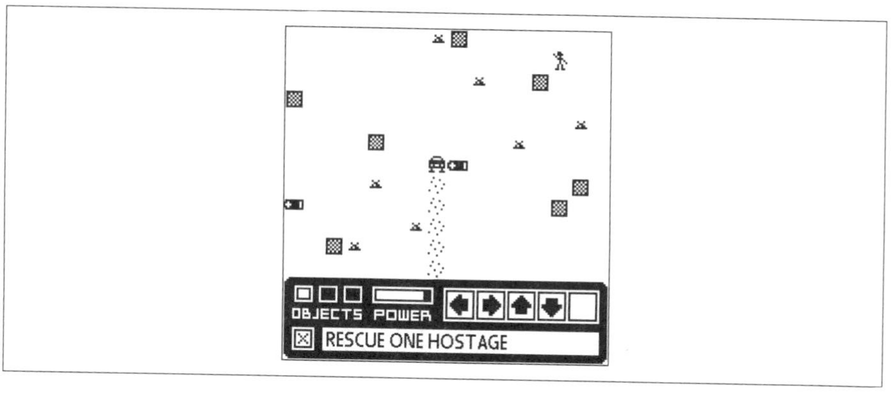

Okay, you got us—it's really just a fancy version of Minesweeper, a silly little game that comes with every version of Windows. This program takes it to a whole new level, though, and the premise really does involve driving a robot around—so give it a shot.

Iron Strategy

Some of the best strategy games have involved programming virtual robots. Another entry in this long and distinguished lineage is Iron Strategy, which, while not groundbreaking in any fundamental way, is a fun way to pass the time with a friend. With each of you controlling a side, you take turns maneuvering robots into position to defeat the other side. Each robot in your small army can move a limited number of spaces, fire a weapon, or perform some other action such as effect repairs to another unit.

There's a real sense of strategy in the game, since you need to eliminate all of the other robots or blow up the enemy base, all without sacrificing your own resources. Most robots are limited to just two or three squares of movement in each turn, and since the game is turn-based, you don't know exactly what your opponent will do based on your own movements. The downside? You can't play against the computer, so it's fun only if you play against someone else.

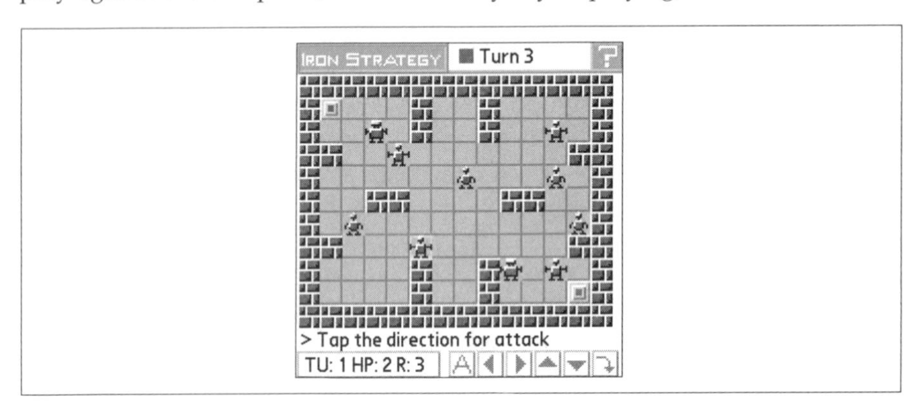

On the PC

In addition to all of the robot-oriented apps on the Palm, you can take part in robot goodness on your desktop computer as well. Depending upon how much of a programmer you liken yourself to be, you might want to try either AT-Robots 3 or MindRover.

AT-Robots3

AT-Robots 3 is a Windows application that lets you design, build, and battle virtual robots based on a simplified version of assembly language. Interested? You can find the program (it's free) at the author's web page, www.nw.fmph.uniba.sk/~9bartelt, or from Dave's web site at www.bydavejohnson.com.

If you download it from Dave's web site, you can find the appropriate link in the Books section.

At its core, AT-Robots 3 is a programming game. You need to create your robot, optimizing the design based on such factors as weapon strength, power demands, shielding, and movement capabilities. All AT-Robots can have parts, including a hull, weapon, scanner, and other components. Designing a robot is pretty straightforward; if you have any programming experience, you can whip up a simple robot in literally minutes. Here's a robot in the process of being created:

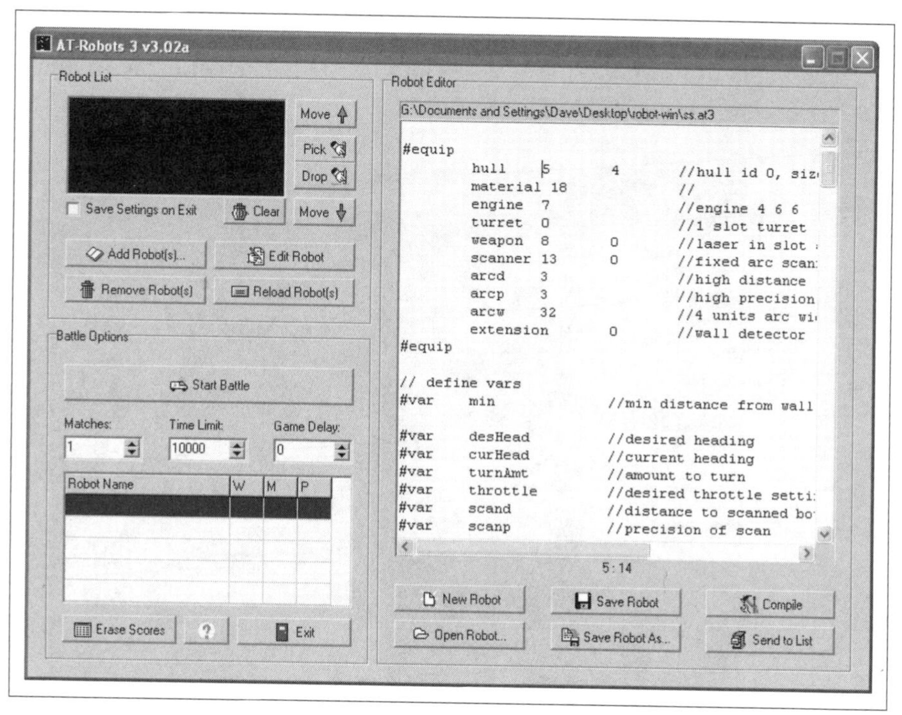

As you can see from the illustration, the robots are little more than scripts, each line of which defines specific robot parameters, like this:

```
#equip
        hull        0     4
        material    5
        engine      7
```

```
        turret      1            //2 slot turret, slots at 0 and 128
        weapon      8     0      //laser in slot at 0
        weapon      8     1      //another laser in slot at 1
#equip
```

The program is well documented and comes with tables that describe all the possible values for the various components; weapon 8, for instance, is a laser, as you can see from the program comments.

After you design your robot, you can unleash it in virtual combat with other robots online. Of course, you need to program the robot's maneuvering logic, but that's half the fun! If you want more insight into programming and competing with AT-Robots 3, you might want to check out Dave's other robot book, *Robot Invasion: 7 Cool and Easy Robot Projects*, from McGraw-Hill/Osborne.

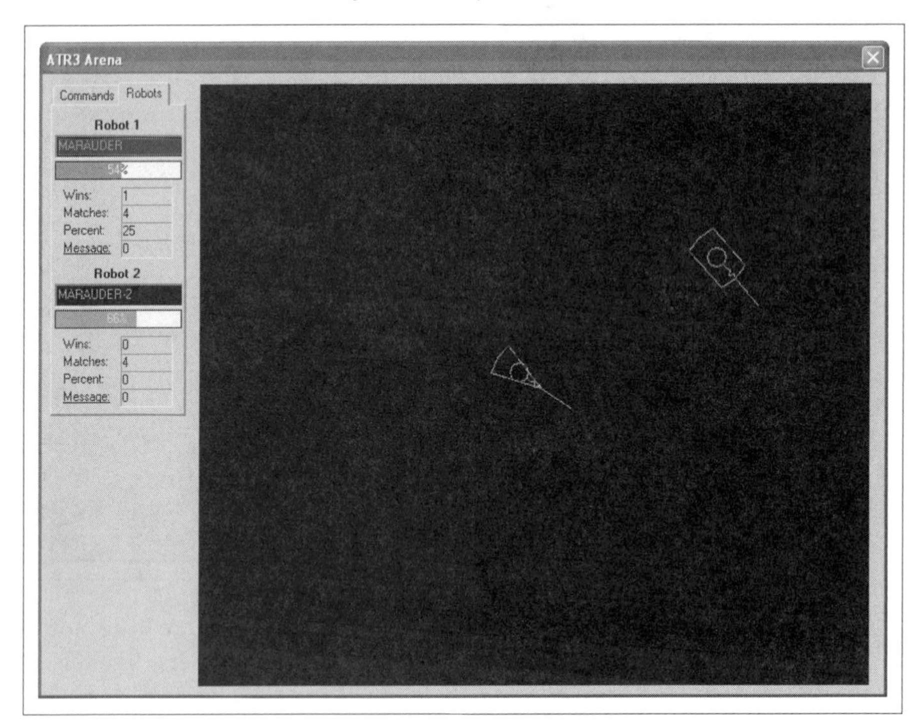

MindRover

If AT-Robots is a bit too imposing for you, you might want to try MindRover instead. Without a doubt the best virtual robot programming game around, MindRover is a lot more "friendly" for the casual roboticist. You can download a trial version from www.mindrover.com, and the full version is available online or in stores for $25.

In MindRover, you get to design robots simply by dragging and dropping components such as sensors, motors, and weapons onto available slots on the robot chassis. Then you program the robot by dragging and dropping logic circuits into place.

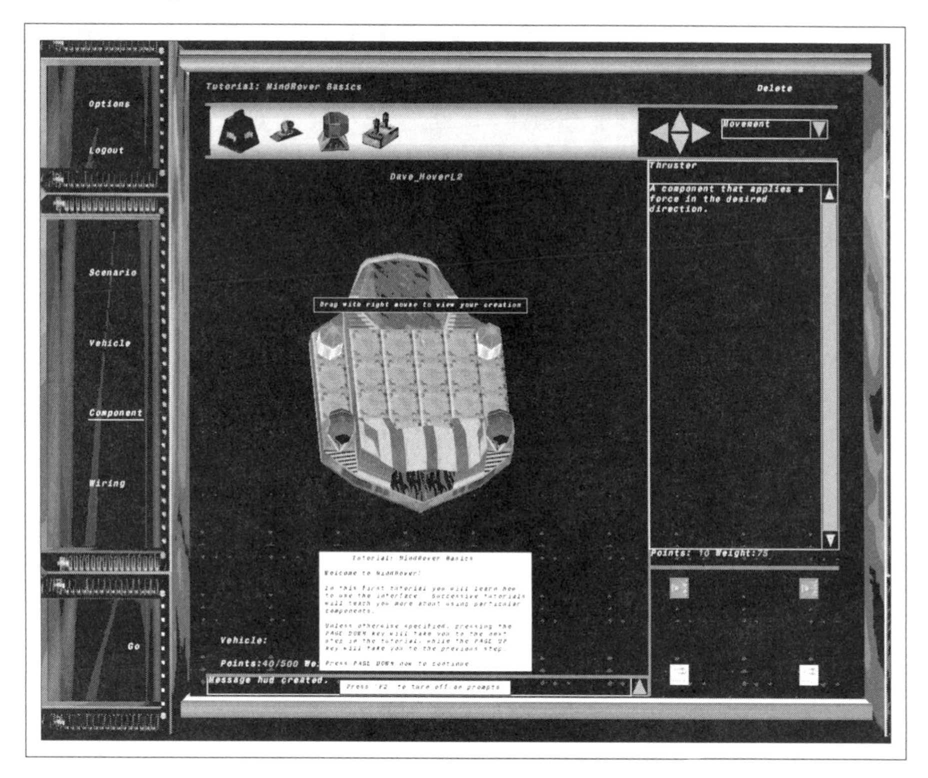

TIP: The MindRover RCX Pack Add-On is a $15 accessory to MindRover. It lets you program Lego Mindstorms robots with the MindRover interface, so you can see if real robots behave the same way as the on-screen bots.

To program the robot, you create "circuits" by dragging connections between the various parts of the robot. To turn, for instance, wire the thruster directly to the sensor, so it'll run whenever it sees an enemy robot.

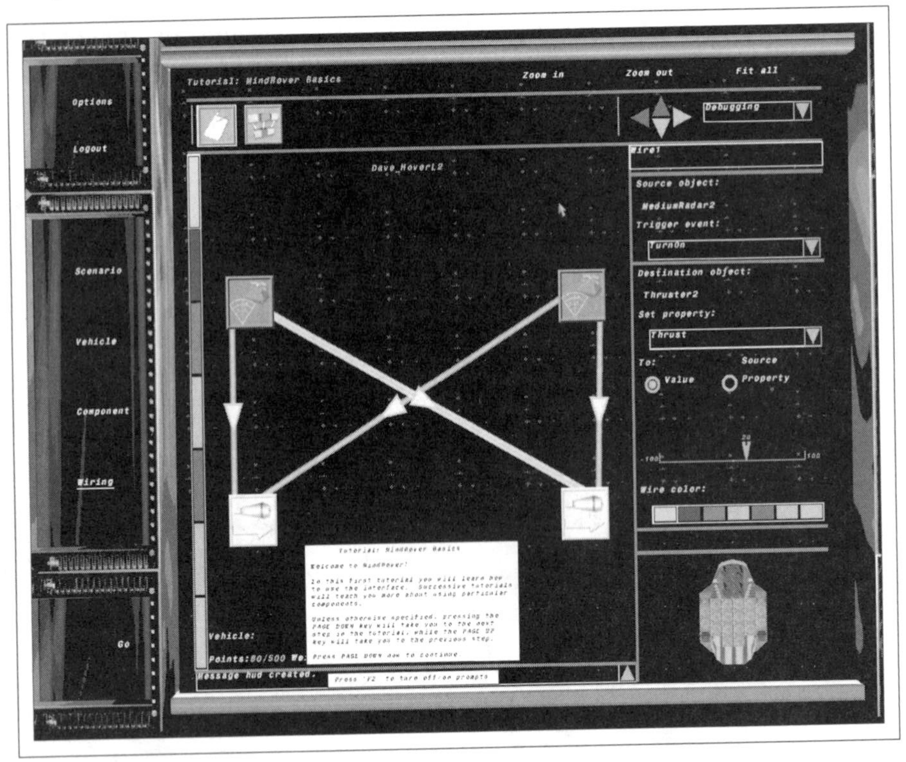

Other tasks require you to use *logic circuits*—you might only fire the rockets if you can see the enemy with both sensors, for instance. That would use an AND circuit.

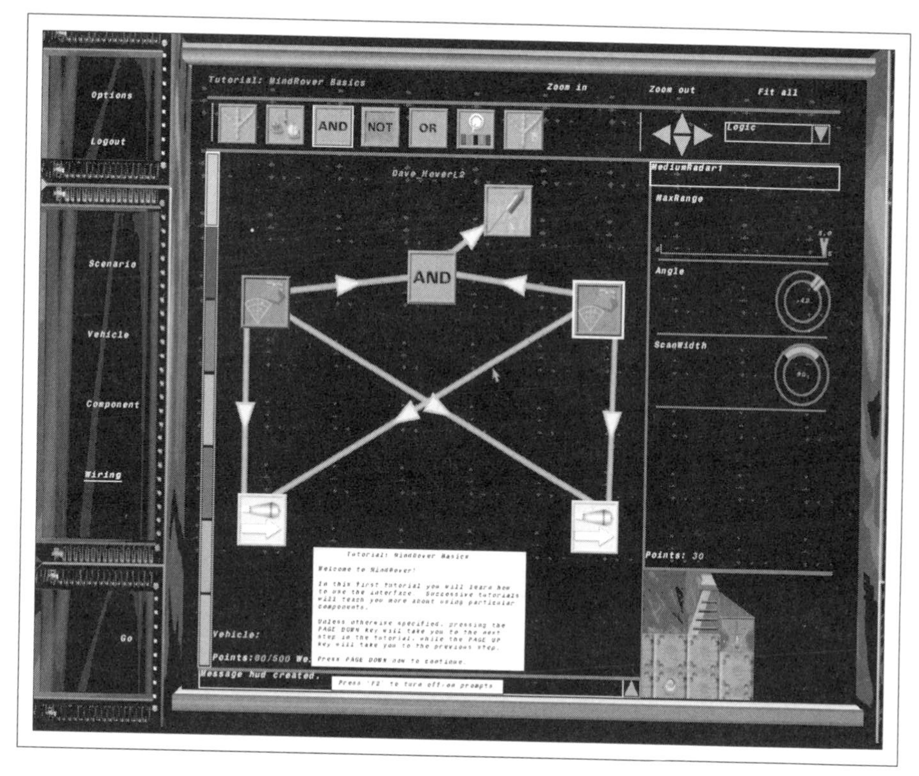

When you're done programming your robot, you can turn it loose in a 3D environment that looks like a living room.

Where To Go From Here

Where to, indeed.

Our journey together is over. Over the course of the past ten chapters, we've shown you how to build a robot, how to program it, and how to modify your robot with new capabilities.

It's now up to you to choose your path.

Are you interested in the hardware side of things? Check the web for resources and information about robots. The PPRK design is unique in the world of wheeled robots, since almost every other wheeled robot uses two or four wheels configured like a cart or car. You may want to explore other wheeled robots or further investigate the capabilities of the three-wheeled design of the PPRK. Beyond wheeled robots, you might branch out to walking robots.

Intrigued by different gadgets? You can look at other sensors or output devices beyond what we've explored in this chapter. What about sound or touch? Can you add a device that makes the robot sensitive to particular sound frequencies? Can you add a sensor that signals the processor when it touches an object? What other devices can you control through the controller board? You may want to look at adding a remote video capability to the robot, so that the robot sends back a picture of where it is. Can you devise a robot that can play a ukulele or a saxophone?

Perhaps you're a software hacker. We've looked at some simple programs for the robot. These programs can provide some interesting behavior, and yet that behavior is somewhat limited. Wall following is useful for the robot, but by itself it does not give the robot a long-term usefulness. Start thinking about the bigger problems you might want your robot to handle and how you could program the robot to handle them.

If you have a BrainStem, you might want to look further into reflexes. While there is a lot of debate about self-awareness and animal intelligence, it seems clear to us that regardless of where one draws the line, many behaviors are driven by reflex and instinct. The less intelligent a creature, the more its behavior is driven be reflex and instinct. Much of artificial intelligence research is directed toward re-creating a complex intelligence. Could the use of reflex programming provide a different approach to artificial intelligence? Fire up that code editor and start programming reflexes.

Regardless of where you go from here, keep your fellow roboticists informed of what you are doing. This book would not have come about if many different people had not shared their work on the web. If you come up with a cool use for the robot or some new techniques for building, or you use some unique sensor, share your work on your own web site. We'd love to hear from you; emails for both Kevin and Dave are listed in the introduction to this book. You can also send pictures and descriptions to www.acroname.com; a gallery on the web site shows a number of projects. Who knows? You could be the next great robotics genius.

Good luck, and have fun!

Appendix

About the CD

What's on the CD

We've covered an awful lot of software in this book. To make it all easier to handle, we've loaded it all onto the CD-ROM that comes with this book. That way, you don't need to spend hours and hours online, going to different Web sites, and using your precious online time to download everything that you need.

You'll find that the CD-ROM is organized into three main folders:

- **Palm Robot Reader** This folder has versions of the Palm Robot Programmer application (from Chapter 6) for a variety of operating environments, including Windows, Mac, and Linux.
- **Other Applications** Here, you'll find all the utilities and other applications we wrote about in the book. Each program has its own folder, making it all easy to find.
- **Licenses** We've rounded up all of the software licenses and stored them in one place. Please be sure to read these before using any program contained on the CD.

How to Use the CD

Using the CD is pretty straightforward. If you want to install an application like the Palm Robot Programmer, do this:

1. Insert the disc in your CD-ROM drive. Depending upon your operating system, the disc will automatically open on your desktop or you'll need to open it yourself.
2. Locate the folder you want to open and double-click on it. If you want to install the Palm Robot Programmer, double-click that folder (Mac users, of course, only need to single-click).
3. Now locate the folder for your operating system and double-click.
4. Double-click the Install icon for the Palm Robot Programmer and follow the onscreen instructions.

That's the basic procedure for installing many of the applications on the CD-ROM. Note that for some apps in the Other Applications folder, there's only one version provided—in most cases, for the Palm OS.

The Palm Robot Programmer Folder

This folder contains all of the installation files for the Palm Robot Programmer software that we developed especially for this book. Installers are included for all major platforms. Note: You must have Java 1.4 installed on your computer to use the Palm Robot Programmer. If you don't already have it, files on the CD will help you get it installed.

To find the right installer for your computer, navigate to the subfolder that best describes your system, such as Windows or Mac. Use this guide to figure out which version to install:

- If you already have Java 1.4 installed on your PC, you can use the NoVM installer, since you don't need to install Java 1.4.
- If you need Java 1.4, look for the VM folder and use that installation. It will install Java 1.4 to your system.
- If the folder for your system only has a NoVM installer, you will have to get Java 1.4 from some other source, such as http://java.sun.com/—we could not provide one for you (sorry!).

Launch the installer and follow the instructions to install the Palm Robot Programmer. You should be able to use the default choices offered by the installer. After installation is complete, locate the program icon that was created by the installer and use it to launch the Palm Robot Programmer.

You can find detailed information on how to create programs with the Palm Robot Programmer in Chapter 6.

Inside the Other Applications Folder

The Other Applications folder is where you'll find, well, everything that isn't the Palm Robot Programmer app. It has Palm OS software, BrainStem software, and other goodies.

You'll find a number of tools, for instance, that you can use to write programs for the PPRK:

- **HotPaw** The HotPaw folder contains the HotPaw Basic interpreter for your Palm OS device. Unpack the archive and load the ybasic.prc and mathlib.prc files to your Palm OS device. You will now be able to run BASIC programs on your PDA. See Chapter 7 for more information on how to write HotPaw Basic programs.

- **MathLib** The MathLib folder contains, not surprisingly, the math library. This library of functions is used by several of the other applications and programs on the CD-ROM. Every application that needs mathlib already includes it as part of its set of files, so you will probably not need to use this folder.
- **NS Basic** This folder contains the demo version of NS Basic that was used in Chapter 7. NS Basic provides a programming environment in which you can create Palm OS programs in a visual environment. In other words, you manipulate forms, buttons, and other user interface elements to create the visual form of your program, and then add the code to give those elements life. NS Basic runs on the Windows platform and has its own installation program. Simply run the installer and follow the prompts to install NS Basic to your computer. You can find more information on how to use NS Basic to create a program in Chapter 7 and in the NS Basic program documentation.
- **PocketC** PocketC is a C interpreter that runs on the Palm OS platform. Simply unzip the archive and load the .prc and .pdb files to your Palm OS device; after that, you will be able to run C programs on your Palm PDA. Information on coding PocketC programs can be found in Chapter 7 and in the PocketC documentation.
- **PtoolBox** This directory contains the PocketToolbox library of functions used with PocketC and the PocketPPRK program.

You'll also find folders that contain some of the programs written in Chapter 5 through Chapter 8:

- **SimpleJig** Contains versions of the Simple Jig program that was built in Chapter 7. It includes versions in BASIC for HotPaw Basic and NS Basic, as well as a C version for PocketC.
- **ToFro** Contains various versions of the ToFro program. There is a version written in TEA (tofor.leaf) for the BrainStem reflex mode, a version in Java from Chapter 8, and a version that can be loaded into the Palm Robot Programmer from Chapter 6.
- **Chbaud** Contains the chbaud.bas program that was used in Chapter 4 to change the baud rate of the SV203 controller. This program runs under HotPaw Basic.

And we've included three of the programs from Chapter 10:

- **FlyingRobots** Contains the Flying Robots game for Palm OS.
- **Welded** Contains the Welded Warriors game for Palm OS.
- **PBrick** Contains the Programmable Brick Library that provides an interface between the Palm OS and a Lego MindStorms controller.

Working with the BrainStem

The BrainStem folder (in the Other Applications folder) contains all the software developed by Acroname for use with the PPRK. This software is further divided by operating system. Acroname has built versions of their software for all the major platforms, including the Palm OS, Windows, Mac, and Linux.

If you navigate to the directory for your system, you will find archive files containing the BrainStem applications that we used in Chapters 4 through 8. To install the software you simply extract the archive onto your hard drive. Here's some of the software you will find:

- **Config** Contains the Config program that can be used to test the sensors and servos of the robot.
- **Console** Contains the Console application that is used to compile and load TEA programs to the BrainStem.
- **GP** Contains the GP application that can be used to test the A/D, digital, and output ports of the BrainStem.
- **Cdev** Contains files useful for writing C programs for the PPRK.
- **Java** Contains files useful for writing Java programs for the PPRK.
- **Ref** The reference manual for the BrainStem.
- **PPRK_SDK** Contains all the files for TEA programming, including Config, Console, and GP.
- **SDK** Everything above in one convenient archive.

Installing Palm Applications

Many of the other applications in this folder are from Chapter 10. So, after scanning that chapter, you might want to try out some of these Palm OS apps.

To install these apps, you need to copy the program to your hard disk and add the PRC files to your Palm PDA via the Install tool. Do this:

1. Find an app that you want to install in the Other Applications folder and drag it to your computer's desktop. Locate all of the .prc and .pdb files—these are the ones you need to copy to your Palm PDA.
2. Start the Palm desktop by choosing Start | Programs | Palm Desktop | Palm Desktop.
3. Click the Install button on the left side of the Palm desktop screen. The Install tool appears.
4. Click the Add button. You see the Open dialog box for selecting applications.
5. Locate the program you want to install and select it. Click the Open button.
6. With your application displayed in the Install Tool dialog box, click Done.
7. The next time you HotSync, the selected application is installed, usually to the Unfiled category on your Palm. If you don't see it there, look in All—some apps get automatically filed elsewhere, like Games or Main.

Inside the PPRK Folder

The PPRK Programs folder (in the Other Applications folder) contains the software that was introduced in Chapter 5. These are programs that were developed by other hobbyists for playing with the PPRK.

The following three PRC files contain programs that you can load to your Palm OS device just like any other PRC file (see "Installing Palm Applications" earlier in this appendix for details):

- MathLib
- PenFollow
- Robot1
- ServoTest

Note that ServoTest and PenFollow conflict with each other, so don't load both of them to your Palm OS device simultaneously. The folder also contains three HotPaw Basic programs for the robot: sequence_motors_test.bas, square2.bas, and triangle3.bas.

Index

C

D

E

F

G

H

I

J

K

L

M

N

O

P

R

INTERNATIONAL CONTACT INFORMATION

AUSTRALIA
McGraw-Hill Book Company Australia Pty. Ltd.
TEL +61-2-9900-1800
FAX +61-2-9878-8881
http://www.mcgraw-hill.com.au
books-it_sydney@mcgraw-hill.com

CANADA
McGraw-Hill Ryerson Ltd.
TEL +905-430-5000
FAX +905-430-5020
http://www.mcgraw-hill.ca

GREECE, MIDDLE EAST, & AFRICA (Excluding South Africa)
McGraw-Hill Hellas
TEL +30-210-6560-990
TEL +30-210-6560-993
TEL +30-210-6560-994
FAX +30-210-6545-525

MEXICO (Also serving Latin America)
McGraw-Hill Interamericana Editores S.A. de C.V.
TEL +525-117-1583
FAX +525-117-1589
http://www.mcgraw-hill.com.mx
fernando_castellanos@mcgraw-hill.com

SINGAPORE (Serving Asia)
McGraw-Hill Book Company
TEL +65-6863-1580
FAX +65-6862-3354
http://www.mcgraw-hill.com.sg
mghasia@mcgraw-hill.com

SOUTH AFRICA
McGraw-Hill South Africa
TEL +27-11-622-7512
FAX +27-11-622-9045
robyn_swanepoel@mcgraw-hill.com

SPAIN
McGraw-Hill/Interamericana de España, S.A.U.
TEL +34-91-180-3000
FAX +34-91-372-8513
http://www.mcgraw-hill.es
professional@mcgraw-hill.es

UNITED KINGDOM, NORTHERN, EASTERN, & CENTRAL EUROPE
McGraw-Hill Education Europe
TEL +44-1-628-502500
FAX +44-1-628-770224
http://www.mcgraw-hill.co.uk
computing_europe@mcgraw-hill.com

ALL OTHER INQUIRIES Contact:
McGraw-Hill/Osborne
TEL +1-510-420-7700
FAX +1-510-420-7703
http://www.osborne.com
omg_international@mcgraw-hill.com

LICENSE AGREEMENT

THIS DISK CONTAINS "THE PALM ROBOT PROGRAMMER", A SOFTWARE PRODUCT OWNED BY THE McGRAW-HILL COMPANIES, IN ("McGRAW-HILL SOFTWARE") AND PROPRIETARY PRODUCTS AND SOFTWARE OWNED BY McGRAW-HILL'S LICENSORS ("THIRD PART PRODUCT" OR "THIRD PARTY PRODUCTS"). YOUR RIGHT TO USE THE McGRAW-HILL SOFTWARE IS GOVERNED BY THE TERMS AN CONDITIONS OF THIS AGREEMENT. YOUR RIGHT TO USE THE THIRD PARTY PRODUCTS IS GOVERNED BY THE TERMS AND CONDITIO OF THE OTHER LICENSE AGREEMENTS INCLUDED IN THE DISK FOR SUCH THIRD PARTY PRODUCTS.

LICENSE: Throughout this License Agreement, "you" shall mean either the individual or the entity whose agent opens this packag You are granted a limited, non-exclusive and non-transferable license to use the McGraw-Hill Software subject to the terms of th Agreement.

(i) You may make one copy of the McGraw-Hill Software for back-up purposes only and you must maintain an accurate record as the location of the back-up at all times.

PROPRIETARY RIGHTS; RESTRICTIONS ON USE AND TRANSFER: All rights (including but not limited to patent, copyrig and trademark) in and to the McGraw-Hill Software are owned by McGraw-Hill and its licensors. You are the owner of the enclose disc on which the McGraw-Hill Software is recorded. You may not use, copy, decompile, disassemble, reverse engineer, modify, r produce, create derivative works, transmit, distribute, sublicense, store in a database or retrieval system of any kind, rent or transf the McGraw-Hill Software, or any portion thereof, in any form or by any means (including electronically or otherwise); except yo may make one copy of the McGraw-Hill Software for back-up purposes only, you must reproduce the copyright notices, tradema notices, legends and logos of McGraw-Hill and its licensors that appear on the McGraw-Hill Software on such back-up copy and yo must maintain an accurate record as to the location of the back-up copy at all times All rights in the McGraw-Hill Software not e pressly granted herein are reserved by McGraw-Hill and its licensors.

TERM: This License Agreement is effective until terminated. It will terminate if you fail to comply with any term or condition of th License Agreement. Upon termination, you are obligated to return to McGraw-Hill the McGraw-Hill Software together with all co ies thereof and to purge and destroy all copies of the McGraw-Hill Software included in any and all systems, servers and facilities.

DISCLAIMER OF WARRANTY: THE McGRAW-HILL SOFTWARE AND THE BACK-UP COPY OF THE McGRAW-HILL SOFTWARE AR LICENSED "AS IS". McGRAW-HILL, ITS LICENSORS AND THE AUTHORS MAKE NO WARRANTIES, EXPRESS OR IMPLIED, AS T RESULTS TO BE OBTAINED BY ANY PERSON OR ENTITY FROM USE OF THE McGRAW-HILL SOFTWARE AND/OR ANY INFORMATIO OR DATA INCLUDED THEREIN. McGRAW-HILL, ITS LICENSORS AND THE AUTHORS MAKE NO EXPRESS OR IMPLIED WARRANTI OF MERCHANTABILITY OR FITNESS FOR A PARTICULAR PURPOSE OR USE WITH RESPECT TO THE McGRAW-HILL SOFTWAR AND/OR ANY INFORMATION OR DATA INCLUDED THEREIN. IN ADDITION, McGRAW-HILL, ITS LICENSORS AND THE AUTHOR MAKE NO WARRANTY REGARDING THE ACCURACY, ADEQUACY OR COMPLETENESS OF THE McGRAW-HILL SOFTWARE AND/O ANY INFORMATION OR DATA INCLUDED THEREIN. NEITHER McGRAW-HILL, ANY OF ITS LICENSORS, NOR THE AUTHOR WARRANT THAT THE FUNCTIONS CONTAINED IN THE McGRAW-HILL SOFTWARE WILL MEET YOUR REQUIREMENTS OR THAT TH OPERATION OF THE McGRAW-HILL SOFTWARE WILL BE UNINTERRUPTED OR ERROR FREE. YOU ASSUME THE ENTIRE RISK WIT RESPECT TO THE QUALITY AND PERFORMANCE OF THE McGRAW-HILL SOFTWARE.

LIMITED WARRANTY FOR DISC: To the original licensee only, McGraw-Hill warrants that the enclosed disc on which th McGraw-Hill Software is recorded is free from defects in materials and workmanship under normal use and service for a period ninety (90) days from the date of purchase. In the event of a defect in the disc covered by the foregoing warranty, McGraw-Hill will r place the disc.

LIMITATION OF LIABILITY: NEITHER McGRAW-HILL, ITS LICENSORS NOR THE AUTHORS SHALL BE LIABLE FOR ANY INDIREC INCIDENTAL, SPECIAL, PUNITIVE, CONSEQUENTIAL OR SIMILAR DAMAGES, SUCH AS BUT NOT LIMITED TO, LOSS OF ANTICIPATE PROFITS OR BENEFITS, RESULTING FROM THE USE OR INABILITY TO USE THE McGRAW-HILL SOFTWARE EVEN IF ANY OF THEM HA BEEN ADVISED OF THE POSSIBILITY OF SUCH DAMAGES. THIS LIMITATION OF LIABILITY SHALL APPLY TO ALL CLAIMS OR CAUSES C ACTION WHATSOEVER WHETHER SUCH CLAIM OR CAUSE OF ACTION ARISES IN CONTRACT, TORT, OR OTHERWISE. Some states do not a low the exclusion or limitation of indirect, special or consequential damages, so the above limitation may not apply to you.

U.S. GOVERNMENT RESTRICTED RIGHTS. If the McGraw-Hill Software is acquired by or for the U.S. Government then it provided with Restricted Rights. Use, duplication or disclosure by the U.S. Government is subject to the restrictions set forth i FAR 52.227-19. The contractor/manufacturer is The McGraw-Hill Companies, Inc. at 1221 Avenue of the Americas, New Yor New York 10020.

GENERAL: This License Agreement constitutes the entire agreement between the parties relating to the McGraw-Hill Softwar Failure of McGraw-Hill to insist at any time on strict compliance with this License Agreement shall not constitute a waiver of an rights under this License Agreement. This License Agreement shall be construed and governed in accordance with the laws of th State of New York without reference to the choice of law principles thereof. If any provision of this License Agreement is held to b contrary to law, that provision will be enforced to the maximum extent permissible and the remaining provisions will remain i full force and effect.